CORKING THE NUCLEAR GENIE
THE PROMISE OF
LOW ENERGY TRANSMUTATION

EDWARD ESKO
ALEX JACK

The Connecticut River at Bellows Falls, 30 miles upriver from Vermont Yankee

The giant casks of spent nuclear fuel loom over the future like a modern, highly toxic Stonehenge. They bear silent witness to our inability, at least for now, to solve the problem of nuclear waste. They are the sentinels of our ignorance. —Edward Esko

Much progress has taken place in establishing firmly the occurrence of different types of transmutation reactions in a wide variety of LENR configurations. —Mahadeva Srinivasan, Bhabha Atomic Research Centre, Mumbai (retired)

This discovery [low energy transmutation] is a revolutionary event in nuclear science and will bring vast influences to modern thought and civilization. These results are so great that we cannot see at present the effects that will come. —Michio Kushi, macrobiotic philosopher and educator

CORKING THE NUCLEAR GENIE
The Promise of Low Energy Transmutation

Contents

Preface by Alex Jack	5
Introduction by Edward Esko	19
Low Energy Transmutation of Nuclear Waste	31
Ten Transmutation Experiments	48
Preliminary Research on Nuclear Remediation	62
Appearance of Barium in Lithium-Iodine Plasma	69
Appendix A: The QR Power Supply	76
Appendix B: Inside the Cool Fusion Lab	77
Resources	87
About the Authors	88

Corking the Nuclear Genie:
The Promise of Low Energy Transmutation
Copyright © 2014 Edward Esko and Alex Jack
ISBN-13: 978-1493664740
ISBN-10: 1493664743

Published by Amber Waves Press
PO Box 487, Becket MA 01223
(413) 623-0012
coolfusion.org
Printed in the U.S.A.
First Edition

Edward Esko (left) and Alex Jack at the lab in Owl's Head

Preface

The storage of nuclear waste still remains unsolved. Nobody knows what to do with it and nobody wants to have the hazardous material sit in their backyard. —**Worldwatch**

Nuclear waste is probably the single greatest security vulnerability in the United States. —**Kevin Kamps, radioactive waste specialist at Beyond Nuclear**

Considering we have only had 10,000 years of written history, we must realize how long future generations will suffer from the onerous legacy left by present societies. Only a few decades of using nuclear energy leave hazardous nuclear waste for an unimaginable number of future generations.
—**Wolfgang Gruendinger, Foundation for the Rights of Future Generations**

Nuclear Genesis
As a child of the nuclear age I was born in August 1945, just days after the bombing of Hiroshima and Nagasaki. My mother worked as the secretary to one of the top administrators of the Manhattan Project. But because of strict wartime security and the classified nature of the work she didn't know it at the time.

The University of Chicago, on whose campus my parents lived and near where I was born, was the site of the first controlled nuclear reaction several years earlier.

My father, a Unitarian minister, was a civil rights and peace activist. In 1957, as a twelve-year-old child, I accompanied him to a peace conference in Japan. In Hiroshima and Nagasaki, we visited hospitalized children my own age that had survived the atomic bombings but suffered from the continuing effects of radiation poisoning. The experience was deeply moving, and like my parents I resolved to devote my life to world peace and health.

In the early 1960s, our family moved to New York, and my dad served as executive director of the National Committee for a Sane Nuclear Policy. SANE successfully led the campaign for an end to nuclear testing during the Kennedy administration and lobbied for nuclear disarmament.

As a reporter in Vietnam during the mid-1960s, I interviewed pilots on an aircraft carrier off the Gulf of Tonkin who revealed that tactical nuclear weapons were secretly stored on board. In an interview, Thich Tri Quang, the leader of the Buddhists in Vietnam, emphasized to me that the decline and destruction of the natural food supply posed as great a threat to culture and civilization as nuclear war. The Zen master encouraged me to study the relation between diet, health, and peace. The encounter changed my life, and when I returned to America, I gravitated to the budding organic foods movement and to the macrobiotic community.

As editor of the *East West Journal* in the late 1970s, I wrote articles on energy and the environment and interviewed Edward Teller, the physicist known as the father of the hydrogen bomb and the role model for Dr. Strangelove. Though I disagreed with his defense of nuclear weapons and energy, Teller was unfailingly polite and I enjoyed talking with him.

Following the atomic bombings of Japan, all the patients at St. Francis's hospital in Nagasaki survived atomic sickness by eating brown rice, miso soup, umeboshi plums, sea vegetables, and other foods with strong anti-radiation effects. [1] Over the years, I helped publicize this story and medical research in Japan and Canada that documented the protective effects of miso, kombu, and other foods. [2] In 1990, as director of the Kushi Institute, I organized an airlift of several thousand pounds of healthful foods to physicians in the Soviet Union and helped introduce macrobiotics in Moscow and St. Petersburg. The doctors used these foods to treat patients and clean up crews in Chelyabinsk, site of Soviet nuclear weapons production, and in Chernobyl who were suffering from leukemia, brain tumors, and thyroid cancer. On a recent trip to the Balkans, I learned first hand about the destructive effects of DU (depleted uranium) used in the war.

Looking back at these many strands in my life, it is clear that coming up with practical, effective solutions to the problems of nuclear energy is a top priority.

A Challenging Legacy

Miraculously, nuclear deterrence kept the peace during the Cold War. But atomic testing, weapons production, and the peaceful use of nuclear energy for civilian purposes left a daunting legacy:

- Estimated worldwide cancer deaths from nuclear fallout range from a low of 11,000 by the Center for Disease Control and Prevention to millions by independent international tribunes [3]

- During the Cold War, several hundred military, civilian, and patient fatalities were recorded as a result of accidents at nuclear power plants, on nuclear submarines, and in medical centers [4]

- Fatalities from the Chernobyl disaster in Ukraine in 1986 include 64 confirmed deaths as of 2008 and predicted cancer deaths ranging from 4000 by the World Health Organization to 200,000 by Greenpeace and 985,000 by a group of Russian researchers. The site will be uninhabitable for 20,000 years [5]

- In Fukushima, Japan, site of a major nuclear accident following an earthquake and tidal wave in 2010, no deaths have been reported. Estimates of extra cancer deaths begin at about 1000. [6] Three years after the accident, radioactivity at the site still has not been contained (see below)

The major types of nuclear waste include:

- High-level waste – spent fuel from nuclear power plants and research reactors; waste streams from reprocessing

- Intermediate level waste – equipment contaminated with radioactive material; disused industrial and medical sources of radioactivity; uranium mine tailings; food irradiation

- Low level waste – miscellaneous lightly contaminated equipment; used protective gear; disposable laboratory vessels; packaging of radioactive chemical compounds; lightly contaminated liquid waste from laboratory procedures; hospital waste

The primary source of nuclear waste comes from the manufacture of nuclear warheads and bombs and their conversion into fuel for electricity. There are an estimated 17,325 nuclear weapons in the world today.

"Just Keep Driving around - We may come up with a solution yet!"

The recent Fukushima accident shows that little was learned about preventing nuclear contamination in the sixty years since the atomic bombings of Hiroshima and Nagasaki. Nearly three years after the accident, the Japanese site remains contaminated and poses a catastrophic threat:

• Massive leakages of heavily contaminated water, collapsing structures, and spent fuels if exposed to air could release 85 times as much lethal cesium as at Chernobyl [7]

• The 400 tons of fuel in the badly damaged pool at Fukushima could release more than 15,000 times the radiation released at Hiroshima

• Up to a thousand tons of heavily contaminated water is pouring each day through the Fukushima site, contaminating the groundwater and fishing off the Japanese coast

Closer to home, radiation leaks in the United States continue to contribute to public apprehension about nuclear technology. In 2012, a leak was found in one of the 177 underground double-shell waste tanks at the Hanford Nuclear Site in Washington that was thought to be stable and impervious to decay.

In the United States, 63,000 tons of nuclear waste remains exactly where it was created—at the power plants themselves. In many cases, these nuclear reactors are situated close to major population centers. Washington, D.C., Boston, New York City, Philadelphia, and Chicago have reactors within a 50-mile fallout radius.

Other Sources of Radioactivity

• *Uranium tailings* are waste by-product materials that remain from processing raw uranium-bearing ore. Though significantly less radioactive than other forms of uranium, they also include thorium-230, radium-226, radon-222 (radon gas) and decayed polonium-210, as well as lead, arsenic, and other hazardous heavy metals. Vast piles of uranium mill tailings dot the landscape around abandoned mining sites in Colorado, New Mexico, and Utah and are a major point of contention with Native American tribal nations in these regions [8]

• *Low-level depleted uranium* (DU), the main by-product of nuclear weapons and nuclear power plant enrichment, poses a serious risk when it is recycled for use by the military and industry. DU is widely used in armor plating and armor-piercing projectiles and has about 60% the radioactivity of natural uranium-235. Because of its relatively short biological half-life (15 days for the body to eliminate half of its strength), DU is less harmful than many radioactive materials. The World Health Organization says it poses no risk of reproductive, developmental, or carcinogenic effects, but British Army doctors have linked DU with lung, lymph, and brain tumors.

The UK Pensions Appeal Tribunal Service concluded that Gulf War veterans had increased risks of birth defects due to DU exposure. [9] *The International Journal of Environmental Research and Public Health* documented an epidemic of cancer and birth defects in Iraq following widespread use of DU in recent conflicts: "Fallujah is experiencing higher rates of cancer, leukemia, and infant mortality than Hiroshima and Nagasaki did in 1945." [10] A similar wave of chronic illness and birth defects has been reported in the Balkans following NATO bombing of Belgrade during the recent Balkans war

• *Medical waste* also poses a hazard. Cesium-137 used for brachytherapy and radiotherapy has a half-life of 30 years and remains harmful for about 300 years. Cobalt-60, iridium-192, strontium-89, iodine-131, ytterbium-90, and other radioactive isotopes also pose threats. [11] The biomedical waste industry is poorly regulated. In Goiania, Brazil, 4 people died in 1987 when exposed to an old radiotherapy source from an abandoned hospital site in the city. About 112,000 people were examined for radioactive contamination and 249 were found to have significant levels of radioactive material in or on their body. In the cleanup operation, topsoil had to be removed from several sites, and several houses were demolished. The International Atomic Energy Agency called it "one of the world's worst radiological incidents." Illegal dumping of radioactive nuclear waste is a problem in India, South Africa, and other developing countries, especially among poor people scavenging in dumps. In Kolkata, India authorities closed down an illegal biomedical waste dump in Chowbaga that used to sell the waste from large private hospitals and nursing homes in the city as plastic scrap to another trader

• *Food irradiation*, another source of nuclear waste, has been linked to creating carcinogens, including benzene, birth defects, and reduced nutrients in food. Although American consumers have largely rejected this technology, it is widely used in herbs and spices. An incident in Decatur, Georgia where water-soluble cesium-137 leaked into the source storage pool requiring NRC intervention has led to sharp decline of this radioisotope and replacement with cobalt-60 [12]

According to proponents of nuclear energy, one person's lifetime nuclear waste would fit in a Coke can. This statement is technically true but misleading. Once the atomic genie has been released into the environment, it is impossible to put back into the bottle. As the Department of Energy states, there are "millions of gallons of radioactive waste" as well as "thousands of tons of spent nuclear fuel and material" and also "huge quantities of contaminated soil and water." [13]

The DOE has launched an ambitious campaign to clean up all contaminated sites by 2025. In Fernald, Ohio, for example, there are 31 million pounds of uranium product, 2.5 billion pounds of waste, 2.75 million cubic yards of contaminated soil and debris and a 223-acre portion of the underlying Great Miami Aquifer with uranium levels above drinking standards. The United States has 108 similar sites, some encompassing thousands of acres.

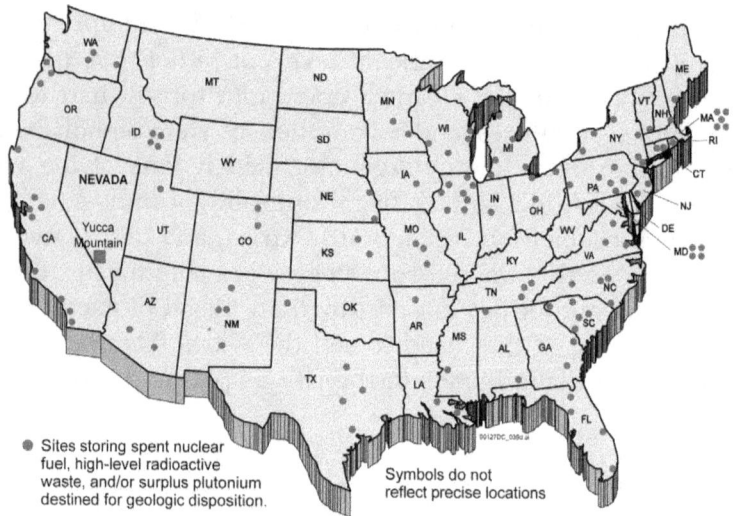

● Sites storing spent nuclear fuel, high-level radioactive waste, and/or surplus plutonium destined for geologic disposition.

Symbols do not reflect precise locations

If Superfund cleanups of ordinary chemicals are any yardstick, cleaning up nuclear sites will be underfunded, haphazard, lengthy, and partial. As the world economy falters, governments have enticed economically depressed areas to house temporary nuclear storage facilities.

Local municipalities are objecting to storing radioactive materials in their backyards. In Cumbria, UK, local authorities voted in 2013 to reject the British government's plan to build an underground storage dump for nuclear waste near the Lake District National Park. Despite sweeteners of hundreds of millions of pounds, the council heeded evidence from independent geologists that "the fractured strata of the county was impossible to entrust with such dangerous material and a hazard lasting millennia." [14]

"Good news! We've been selected as a potential burial site for nuclear waste."

Third World countries are also turning down offers to store nuclear waste. In Mongolia, citizens protested against plans secretly arranged by American, Japanese, and Mongolian government officials to build nuclear waste facilities in Central Asia.

Off the coast of Somalia, local clans secretly hired by Italy's state energy research agency Enea in the 1980s and 1990s have illegally dumped nuclear waste. Such dumping has contributed to the breakdown of civil authority, the rise of piracy, and the sharp health and environmental decline in that troubled region.

A comprehensive litany of nuclear accidents, injuries, and deaths has been documented by Eric Schlosser (author of *Junk Food Nation*) in *Command and Control*, his new book on the illusions of nuclear safety.

Thinking Outside the Nuclear Storage Box
What if cesium-137, strontium-90, uranium-235, and other highly radioactive elements could be converted into harmless, stable isotopes?

Radioactive cesium, the main byproduct of nuclear accidents in Chernobyl and Fukushima, has a half-life of 30 years and will pose a serious health risk to people for about 300 years. (As a rule of thumb, a radioactive substance is hazardous for ten times its half-life, or enough time to reduce its lethality to about 0.1%.) Two of the most long-lived fission products are technetium-99 (half life 220,000 years) and iodine-129 (half life 15.7 million years). Other lethal isotopes are neptunium-237 (half life 2 million years) and plutonium-239 (half life 24,000 years). One of the most long-lived, uranium-235 has a half-life of 700 million years and, applying the rule of ten, will not completely decay for 7 billion years. That is the estimated time the sun will last and hence all future life on earth. Fortunately, its lethality evaporates long before that time, as do other Methuselah-like fission products whose half-lives extend billions of years!

The Cold War and the era of nuclear power have resulted in millions of tons of nuclear waste for which there is no permanent storage. All current solutions, including burying waste in salt mines, are inherently unstable and will inevitably contaminate the planet. But what if, rather than trying to contain or shield these deadly isotopes, radioactive cesium-137 could be converted into the highly sought-after rare earth element neodymium, strontium-90 into ruthenium, the first element in the scarce platinum group of elements, or highly radioactive iodine-129 into a stable and utterly harmless isotope of barium? Or in the case of uranium, which has no nonradioactive form, transformed into bismuth, the last stable element in the Periodic Table? At a stroke, the nuclear genie could be put back in the bottle and useful raw materials produced.

The transmutation of elements, or Cool Fusion, is the goal of Quantum Rabbit (QR), the small macrobiotic company I helped launch with Edward Esko and Woody Johnson. In experiments at laboratories in New England over the last eight years, QR has successfully created small amounts of about 20 key industrial metals, including iron, copper, titanium, silver, gold, aluminum, and scandium, from carbon, silicon, and other common elements. The tests build on pioneer studies conducted by Louis Kervran, a French biochemist, and macrobiotic educators George Ohsawa and Michio Kushi in the 1960s. Michio Kushi continues to teach and write actively in his late '80s and serves as an advisor to QR.

In early 2013, QR conducted its first isotope studies. In vacuum tube experiments in Owl's Head, Maine, we set out to transmute iodine-127 into barium-134 through low energy nuclear reactions (LENR). Lithium-7 was used as a catalyst, producing the formula: iodine-127 + lithium-7 → barium-134.

Barium, a stable element in the center of the Periodic Table, is best known as a contrast for radioactive X-rays in the Barium Enema test. Barium was also in the news this spring when the ratio of barium to calcium in teeth was used to determine how long Neanderthals nursed their children (7 months). As this archeology study demonstrated, the barium remained stable for over 100,000 years.

The QR study involved three separate tests. Upper and lower copper electrodes were inserted into both ends of a vertical vacuum tube (see diagram above). A lithium plug was then inserted in the center of the lower electrodes in Tests 1 and 2. Pure iodine crystals were placed on top the lithium inserts. In Test 3, a small piece of lithium was placed in the center of the lower copper electrode surrounded by iodine crystals instead of the plug.

After pumping down to vacuum, oxygen was admitted and the power turned on. When the arc was established and plasma struck, additional heat was provided by a hand-held torch. The arc was maintained for about 10-15 minutes, at which time the power was disconnected and the tubes allowed to cool. [15]

Residue from each test was collected and sent to Northern Analytical Laboratory in Londonderry, N.H. for independent analysis. As predicted, barium appeared in all three samples, ranging from 3.5 ppm to 1.8 ppm, to 463 ppm. The distribution of barium isotopes in the test samples significantly varied from the distribution in nature, suggesting the possibility that detected barium was newly created through LENR and not from contamination in the laboratory. Barium-134 occurs in only 2.4% of all the barium on the planet. In the QR tests, it measured up to 4.04% or nearly double. Barium-132, which is found at 0.101% in nature appeared up to 3.92%, or about 40 times higher than normal.

Northern Analytical Laboratory was so startled by the anomalous findings that it reevaluated the tests on its own initiative. It confirmed the results but theorized that the spectroscopic analysis of barium overlapped with that of zinc, which has a similar profile, accounting for the discrepancies.

Since there was no zinc used in the experiments, however, the chance of contamination was remote. Most likely, low energy fusion between iodine (atomic number 53) and lithium (atomic number 3) could explain the consistent appearance of barium (atomic number 56) in all tests.

The distribution of barium isotopes can be accounted for by the fusion of iodine-127 and lithium-7. That formula was predicted beforehand. Barium-132 may have arisen after the ejection of a neutron at the moment of fusion between the lithium and iodine.

Such studies need to be replicated and eventually radioactive materials introduced. For that, a university or government nuclear laboratory would need to be involved. And then when proof of concept is definitely established, engineers would need to be brought in to scale the process up for practical use, such as designing on site remediation units that could immediately decontaminate nuclear waste from atomic power plants, factories, and hospitals.

A coordinated effort by small companies like QR, NGOs, industry, the military, and government could creatively solve the nuclear storage crisis within our lifetimes, bequeath a nuclear-free planet to our children and future generations, and cork the nuclear genie once and for all.

Alex Jack
The Berkshires
October 26, 2013

[1] Dr. Tatsuichiro Akizuki, Gordon Honeycombe and Keiichi Nagata, Nagasaki 1945, *Quartet Books*, London, 1982.
[2] The scientific studies are cited in Hiroko Furo, *Healing with Miso*, Amberwaves, 2008.
[3] Rob Edwards, "Nuclear test fall-out killed thousands in US, *New Scientist*, March 2002.
[4] "List of nuclear and radiation accidents by death toll," Wikipedia, accessed October 27, 2013. Also see Alexey V. Yablokov et al, *Chernobyl: Consequences of the Catastrophe for People and the Environment* (2007), which forecasts 985,000 deaths from the Chernobyl accident between 1986 and 2004.
[5] Ibid.
[6] Ibid.
[7] Mae Wan Ho, "The World Must Take Charge at Fukushima," *ISIS Report*, September 30, 2013. Institute of Science in Society. www.i-sis.org.uk.
[8] "Uranium tailings danger to health, environment," Nuclear Australia, nuclearnewsaustralia.wordpress.com.
[9] Mike Ludwig, "Depleted Uranium Contamination is Still Spreading in Iraq," Truth-out.org, March 2013.
[10] C. Busby et al., "Cancer, Infant Mortality and Birth Sex-Ratio in Fallujah, Iraq 2005–2009,"*Int. J. Environ. Res. Public Health*, 2010, 7(7), 2828-2837.
[11] "Goiânia accident," Wikipedia, accessed October 27, 2013.

[12] "Food irradiation," www.sustainabletable.org/728/food-irradiation.
[13] U.S. Department of Energy Environmental Management, "Department of Energy Five Year Plan FY 2007-FY 2011 Volume II." Retrieved 8 April 2007.
[14] Martin Wainwright, "Cambria Rejects Underground Nuclear Storage Dump," *The Guardian*, January 30, 2013.
[15] Real time video of the Iodine experiments can be viewed at YouTube.com/QuantumRabbit. Click on "Cool Fusion Science Iodine Study."

Introduction

The problem is how to keep radioactive waste in storage until it decays after hundreds of thousands of years. The geologic deposit must be absolutely reliable as the quantities of poison are tremendous. It is very difficult to satisfy these requirements for the simple reason that we have had no practical experience with such a long-term project. —**Hannes Alfven, Nobel laureate in physics**

Once each season for several years the cool fusion research team would meet at the Moore Mill building in Bellows Falls, Vermont, to conduct tabletop carbon-arc experiments. Alex Jack and I would drive up from Western Massachusetts. The drive took us east on the Mass Pike and then north up route 91 past Brattleboro to the Rockingham exit. Bellows Falls is on the Connecticut River. Thirty miles downriver, in the town of Vernon, is the Vermont Yankee nuclear power plant.

Vermont Yankee epitomizes the chaotic state of affairs within the nuclear industry. The amount of radioactive waste stored at Vermont Yankee is substantial. It exceeds that of all four damaged reactors at Fukushima in Japan. The used fuel rods at Vermont Yankee are about one million times more radioactive than they were before being used in the reactor. The rods are hot enough to catch fire if they are not stored under water. Five hundred tons of spent fuel is now being stored in pools of water seven stories above ground (Fig.1). [1]

Following decades of conflict with local residents and with the State of Vermont, it was announced that Vermont Yankee would close by the end of 2014. The Yankee plant went on line in 1972 and is the same General Electric boiling water system as the failed nuclear reactors at Fukushima. Following the disaster at Fukushima, Entergy, the owner of Vermont Yankee, was being pressured to make expensive modifications to improve safety.

Fig. 1. The waste storage facility at Vermont Yankee

According to nuclear engineer Arnie Gunderson, "The problem is Fukushima modifications are coming due, that's close to $100 million, plus the leaky condenser, that's another $100 million. So it just made no economic sense." [2] Financial experts point to low natural gas prices as another factor in the closure of the plant, explaining that competition from cheap natural gas-fired electric plants severely limited Vermont Yankee's margins.

Julien Dumoulien-Smith, an energy analyst with UBS financial services predicted, "This is likely a bellwether as far as it goes for the nuclear industry. In some sense, this is the first or perhaps the second of a large wave of potential nuclear power plant retirements in the country." [3]

The closing of the plant means that Vermont Yankee will no longer be producing radioactive waste. However, still unresolved is what will happen with the tons of highly radioactive spent fuel rods stored at the site. Local residents are justifiably concerned. One resident, who is the director of a local citizen advocacy group, stated, "There are at least 530 tons of high-level (radioactive) waste, which we've said still needs to come out of that spent fuel pool. This isn't over. The struggle is now about cleanup." [4]

Another resident stated, "The banks of the river, within a 500-year flood plain, is not the best place to store high-level radioactive waste for even a short period of time. Unfortunately, there is no long-term storage in the U.S., and we're probably stuck with it there, just like at Yankee Rowe." [5] Yankee Rowe is the former nuclear plant in nearby Rowe, Massachusetts that was shut down in 1992. High-level radioactive waste from the plant is now in temporary storage in 16 dry casks at the site (Fig. 2). The giant casks of spent nuclear fuel loom over the future like a modern, highly toxic Stonehenge. They bear silent witness to our inability, at least for now, to solve the problem of nuclear waste. They are the sentinels of our ignorance.

Fig. 2. Dry casks at Yankee Rowe

A third resident expressed concern about the safety of Vermont Yankee's phase-out: "Now until the fall of 2014 will be the most dangerous year of their operation of the plant, because the plant will be older than ever, parts will be more brittle than ever, they will be more reluctant than ever to repair and replace parts ... and workers who have already been let go are going to be leaving in much larger numbers. To expect that Entergy is going to be taking every single precaution to keep the plant as safe as it possibly can be is unfortunately unrealistic." [6]

A member of the citizen advocacy group, the New England Coalition, explained the situation at Vermont Yankee quite succinctly: "One fundamental purpose of our advocacy has always been to try to protect the public and the environment from nuclear waste—waste in the fuel, in the reactor, in the pool, out-in-the-yard, soon to be released in the next reactor or fuel handling accident, and out on the wind. Soon, Entergy Vermont Yankee, a nuclear waste pile that generated electricity will stop generating electricity—and it will either be mothballed or promptly torn apart, but it will be, absent electricity generation, just a nuclear waste pile ... from which the public and the environment need to be protected."[7]

On a planetary scale, there are more than 430 locations around the world where nuclear waste continues to accumulate. Most is stored at individual reactor sites. Nuclear reactors on planet earth create about 10,000 metric tons of spent nuclear fuel each year. Thus, the problem gets worse with each passing day. The shutdown and decommission of nuclear power plants solves only the problem of new nuclear waste. It does nothing to solve the problem of already existing waste. Moreover, the process is expensive (between $300 million to $5.6 billion per unit), time-consuming, and hazardous to workers and the natural environment. It opens a window for disaster caused by human error, accident, or sabotage.

In the U.S. there are 13 reactors that have shut down and are in the process of decommission. None have fully completed the process that can last as long as 100 years. The timeframes when dealing with nuclear waste are enormous; they range from 10,000 years to millions of years. Storage of nuclear waste, whether "temporary" or "permanent," does not solve the problem. It simply passes it on to future generations. The very existence of nuclear waste is itself the problem. For the sake of future generations, we need to seriously investigate promising ideas not just for storing nuclear waste, but also for actually *getting rid* of it.

Toward that end, Quantum Rabbit LLC has been conducting tabletop research on the low energy transmutation. Transmutation is defined as transforming one atom into another by changing its nuclear structure. Low energy transmutation attempts to achieve this with simple tabletop equipment, using electric power from car batteries, solar panels, a HUBERT® portable generator, and the wall socket, with relatively low temperatures and low pressures created in simple glass vacuum tubes. This is in contrast to the high-energy accelerator transmutation of waste (ATW) being studied around the world as a possible solution to the radioactive waste problem.

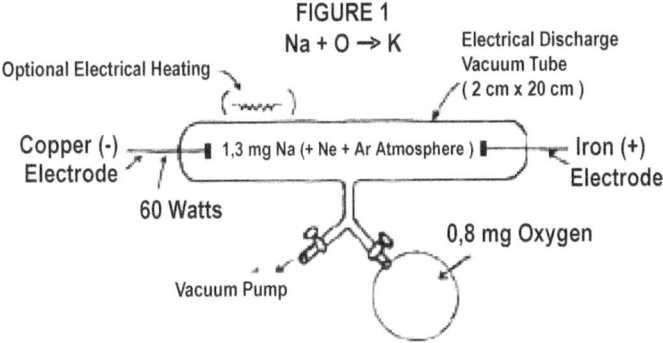

Fig. 3. The Ohsawa sodium into potassium experiment

The Quantum Rabbit experiments are based on the work of George Ohsawa, the Japanese philosopher and educator, who, together with French biochemist Louis Kervran, developed the theory of biological and elemental transmutation. In 1964, Ohsawa claimed to have transmuted sodium into potassium in a tabletop experiment conducted in Tokyo. [8] Ohsawa claimed that this transmutation took place through a process of low energy fusion in which two lighter elements, sodium and oxygen, fused to form the heavier element potassium. A diagram of his experiment is shown in Fig. 3.

Sodium has the atomic number 11 and the atomic weight 23. Oxygen has the atomic number 8 and the atomic weight 16. If we add the atomic number and weight of sodium with the atomic number and weight of oxygen, we get the atomic number (19) and weight (39) of potassium.

The formula is written as follows, with the plus sign (+) indicating a fusion reaction and the arrow (→) indicating the fusion product:

$$^{11}Na_{23} + {^8}O_{16} \rightarrow {^{19}}K_{39}$$

Following this initial experiment, Ohsawa began researching the transmutation of carbon into iron, with the addition of oxygen. In Cambridge, Massachusetts, Michio Kushi, a student of Ohsawa, performed experiments in which carbon and oxygen were apparently fused to form iron. [9] The year was 1965. The carbon into iron experiment was done in the open air and is another example of low energy fusion. Graphite (carbon) powder was placed on a copper plate. A graphite rod was connected to several car batteries. An arc was struck when the rod was placed near the graphite powder. The arc sucked in oxygen and nitrogen from the surrounding air. The process was repeated and the remaining powder was found to be magnetic and to contain trace amounts of iron. The low energy fusion formula can be written as follows:

$$2({^6}C_{12} + {^8}O_{16}) \rightarrow {^{26}}Fe_{56} + 2\text{ protons}$$

Forty years later, in 2004, I joined with Alex Jack and Woody Johnson to form Quantum Rabbit LLC (QR), a Massachusetts company, to develop the ideas and experiments of Ohsawa and Kushi. We established a small lab in Nashua, New Hampshire to begin experiments on low energy transmutation using custom designed glass vacuum tubes. Carbon arc experiments were also started at Woodland Energy in Bellows Falls, Vermont. In 2009, the vacuum lab moved to Owls Head, Maine.

Table 1. Results of Carbon Arc Experiments

Element	Concentration (ppm*)
Silicon	10,500
Magnesium	1800
Iron	4700
Aluminum	7800
Titanium	440
Scandium	35
Cobalt	160
Nickel	1120

*Parts per million

As with the experiments conducted forty years earlier, in our experiments, when subjected to the arcing process, carbon displayed magnetic activity. Placing the treated graphite powder on white paper and waving a neodymium magnet underneath it repeatedly confirmed magnetic properties. [10] Not only was the graphite magnetic, but also when analyzed independently, it showed traces of the elements shown in Table 1. [11] Low energy transmutation may help explain the presence of anomalous elements in treated graphite (Table 2).

Table 2. Possible Low Energy Fusion Reactions in Treated Graphite*

Magnesium-24 $^{12}C + {}^{12}C \rightarrow {}^{24}Mg$	Calcium-40 $^{24}Mg + {}^{16}O \rightarrow {}^{40}Ca$
Aluminum-27 $^{12}C + {}^{15}N \rightarrow {}^{27}Al$	Titanium-44 $^{28}Si + {}^{16}O \rightarrow {}^{44}Ti$
Silicon-28 $^{12}C + {}^{16}O \rightarrow {}^{28}Si$	Iron-56 $2({}^{12}C + {}^{16}O) \rightarrow {}^{56}Fe$ - 2 protons
Potassium-39 $^{26}Mg + {}^{14}N \rightarrow {}^{39}K$	Scandium-45 $^{30}Si + {}^{15}N \rightarrow {}^{45}Sc$

*The gases involved in these reactions, oxygen (O) and nitrogen (N), are from the atmosphere. Elements shown with their atomic weights

Following a vacuum arc experiment in Nashua on May 30, 2008, 1,500-ppm copper was unexpectedly detected on an electrode made of stainless steel. Aside from stainless electrodes, lithium and oxygen were the only elements introduced into the experiment, suggesting the possibility that the copper was newly created through transmutation. The stainless electrodes were composed mostly of iron (Fe). Low energy fusion of the iron in the electrodes with the lithium (Li) test material could explain the appearance of copper:

$$^{56}Fe + {}^{7}Li \rightarrow {}^{63}Cu$$

The test was repeated on Dec. 30, 2008 with similar results. In the Dec. 30 test, copper was found on the electrodes at 557 ppm. [12] These results appeared to confirm the theory of low energy transmutation and prompted Quantum Rabbit to initiate a series of metal vapor tests using a variety of electrode elements and catalyst materials.

The metal vapor experiments are described in our book, *Cool Fusion* (Amber Waves, 2012). The results of the metal vapor experiments are summarized in Table 3.

Table 3. Results of Vacuum Arc Experiments

Element	Concentration (ppm*)	Catalyst	Electrode
Copper	1,500	Lithium	Stainless
Germanium	3,570	Lithium	Copper
Tin	5	Lithium	Silver
Potassium	750	Lithium/sulfur	Copper
Palladium	181	Sulfur	Zinc
Strontium	3.5	Oxygen	Zinc
Chromium	10	Sulfur/oxygen	Copper
Aluminum	342	Boron/oxygen	Copper
Scandium	18	Boron/sulfur	Copper
Selenium	12	Boron	Copper

*Parts per million

In every case, the results were predicted beforehand. Formulas based on the theory of low energy transmutation guided the protocol for each experiment. Test samples were carefully collected and sent to an outside lab for analysis by ICP (Inductively Coupled Plasma Spectroscopy). Low energy fusion formulas explaining the vacuum arc anomalies are presented in Table 4.

Table 4. Possible Low Energy Fusion Reactions in Vacuum Arc Experiments*

Copper-63	Strontium-86
$^{56}Fe + {}^{7}Li \rightarrow {}^{63}Cu$	$^{68}Zn + {}^{18}O \rightarrow {}^{86}Sr$
Germanium-72	Chromium-50
$^{65}Cu + {}^{7}Li \rightarrow {}^{72}Ge$	$^{34}S + {}^{16}O \rightarrow {}^{50}Cr$
Tin-116	Aluminum-27
$^{109}Ag + {}^{7}Li \rightarrow {}^{116}Sn$	$^{16}O + {}^{11}B \rightarrow {}^{27}Al$
Potassium-39	Scandium-45
$^{32}S + {}^{7}Li \rightarrow {}^{39}K$	$^{34}S + {}^{11}B \rightarrow {}^{45}Sc$
Palladium-102	Selenium-76
$^{68}Zn + {}^{34}S \rightarrow {}^{102}Pd$	$^{65}Cu + {}^{11}B \rightarrow {}^{76}Se$

*Elements shown with their atomic weights

In 2005 I began working on formulas for the low energy fission of heavy elements. Low energy fission is the opposite of low energy fusion. In low energy fission, a heavy element, such as lead or bismuth, splits into two lighter elements. In my research, I noted textural and energetic similarities between the elements lead, gold, and lithium. I developed the following formula based on these observations:

$$^{204}Pb \rightarrow {}^{197}Au + {}^{7}Li$$

The first sketch I did of the experiment is shown in Fig. 4. In it I proposed initiating the low energy fission of lead into gold by a low energy fusion reaction in which lithium fuses with oxygen to form sodium. The key to achieving this transmutation is to pull atoms of lithium out of atoms of lead. Michio Kushi refers to this process as "nuclear subtraction." As he states in *The Philosopher's Stone*, "After studying this problem, I began to see several possible means for transmuting gold using the addition method. But unless I could understand the subtraction method, my understanding would have remained incomplete." [13]

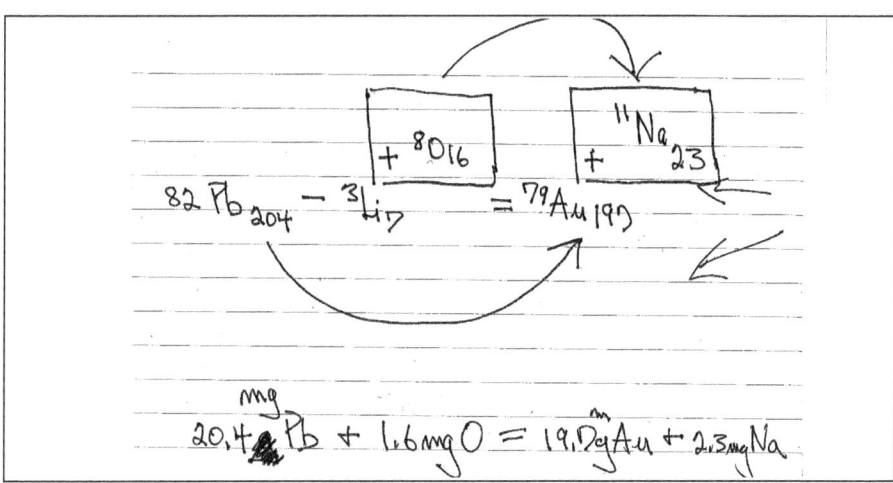

Figure 4. July 2005 sketch of the Pb → Au experiment

After doing experiments with lithium, sodium, oxygen, and sulfur, I started to think that it would be easier for lithium to react with solid sulfur rather than with gaseous oxygen.

In the theory of low energy transmutation, sulfur is a form of crystalized oxygen. It is formed by the fusion of two oxygen atoms ($^{16}O + ^{16}O \rightarrow ^{32}S$). So when it became time to do the actual experiment, we decided to go with sulfur, with the goal being to trigger the low energy fission of lead into gold through the following reaction:

$$^7Li + ^{32}S \rightarrow ^{39}K$$
Lithium-7 + sulfur-32 → potassium-39

I hypothesized that the low energy fusion of lithium and sulfur would release the energy needed to fission lead into the two lighter elements. That is why, in each of the experiments performed to test this hypothesis, in addition to gold, our analysis tested for potassium. The presence of potassium could provide evidence that at least one of the predicted low energy nuclear reactions took place.

The QR team tested the Pb → Au formula over a four-year period, beginning in 2008. Lab sessions took place in Nashua on Dec. 30, 2008 and Owls Head on July 30, 2009, Sept. 27, 2011, and April 11, 2012. These trials are described in my paper, *Ten Transmutation Experiments*. Results of the experiments were mixed. Potassium appeared in all the experiments, ranging in concentration from 4.41 ppm to 750 ppm. Gold was present in the first five experiments, ranging in concentration from 5 ppm to 5540 ppm, but not in the second five tests. In these later tests, held in Owls Head on April 10, 2012, the value for gold following ICP analysis was consistently <0.0 ppm, or below the detection limit.

Nevertheless, these results led me to speculate about the possibility of using low energy methods to transmute the radioactive isotopes found in nuclear waste into stable non-radioactive elements. In theory, the process of subtracting lithium or another light element from super-heavy elements could be used to condense the decay cycle of radioactive elements such as uranium-235 and plutonium-239 from thousands or millions of years to a few years at most. Proposed experiments are outlined in my paper, *Low Energy Transmutation of Nuclear Waste*.

On a parallel track, low energy fusion could also be deployed to instantly convert radioactive fission products like iodine-129, technetium-99, and cesium-137 into useful elements like barium, palladium, and neodymium. Proposals for transmutation of these radioactive waste products are also presented in my paper.

On April 10, 2013, QR conducted the first of what is hoped will be a series of preliminary tests on the remediation of nuclear waste. The test involved the transmutation of stable iodine-127 into barium-134 through a simple low energy fusion reaction. Radioactive iodine-129 is created by the fission of uranium and plutonium in nuclear reactors. With a half-life of 15.7 million years, iodine-129 poses a significant threat to human health and the environment. Successful transmutation of stable iodine-127 could provide a pathway for research on converting radioactive iodine-129 into non-radioactive barium. The 2013 experiment tested the formula:

$$^{127}I + {}^{7}Li \rightarrow {}^{134}Ba$$

As predicted, barium appeared in all three test samples, ranging as high as 463 ppm from a starting concentration (in the iodine used in the test) of 0.43 ppm, more than a thousand-fold increase. Moreover, the distribution of isotopes in the treated sample varied considerably from the distribution of isotopes found in nature. Novel isotope distribution is considered the gold standard for ruling out contamination and proving transmutation. The experiments are presented in detail in my paper, *Appearance of Barium in Lithium-Iodine Plasma*.

Clearly, much work needs to be done. We plan to continue experiments at our small lab in Maine, but with limited time and resources, our work remains preliminary, like that of Steve Jobs and Steve Wozniak in their garage in Silicon Valley or the Wright Brothers in their bicycle shop in Ohio. Our hope is to eventually partner with established labs or universities to take our preliminary work to the next level.

Nuclear energy was unleashed seventy years ago with the Manhattan Project. In today's dollars, the creation of nuclear power cost about $26-billion. With research and development in 30 locations, the project employed 130,000 people. It may require a similar effort today, with researchers around the globe committed to the same goal, to put the nuclear genie back into the bottle. We owe it to future generations to at least try.

Edward Esko
Pittsfield, Mass. and Lilac Park, Lenox, Mass.
Autumn Equinox, 2013

[1] Vermont Citizens Action Network, "Where does Vermont Yankee's waste go?" vtcitizens.org/waste.
[2] Dillon, John, "Citing Economics, Entergy to Close Vermont Yankee by End of 2014," digitial.vpr.net, August 27, 2013.
[3] Ibid.
[4] "Anti-nuclear community celebrates Vermont Yankee closing," timesargus.com, September 2, 2013.
[5] Ibid.
[6] Ibid.
[7] Ibid.
[8] Kushi, Michio, *The Philosopher's Stone*, One Peaceful World Press, Becket, Mass., USA, pp. 23-28, 1994.
[9] Kushi, Michio, *The Philosopher's Stone*, One Peaceful World Press, Becket, Mass., USA, pp. 28-33, 1994.
[10] Esko, Edward and Jack, Alex, *Cool Fusion*, second edition, Amber Waves, Becket, Mass. USA, pp. 56-60, 2012.
[11] Ibid.
[12] Esko, Edward and Jack, Alex, *Cool Fusion*, second edition, Amber Waves, Becket, Mass. USA, pp. 65-70, 2012.
[13] Kushi, Michio, *The Philosopher's Stone*, One Peaceful World Press, Becket, Mass., USA, pp. 33-36, 1994.

Low Energy Transmutation of Nuclear Waste

Abstract
Quantum Rabbit (QR) research on the low energy fusion and fission (low energy nuclear reactions, or LENR) of various elements indicates possible pathways for applying that process to reducing nuclear materials. In a New Energy Foundation (NEF)-funded test conducted at Quantum Rabbit lab in Owls Head, ME, QR researchers initiated a possible low energy fission reaction in which lead-204 fissioned into lithium and gold ($^{204}Pb \rightarrow {}^7Li + {}^{197}Au$). [1] This reaction may have been triggered by a low energy fusion reaction in which lithium fused with sulfur to form potassium ($^7Li + {}^{32}S \rightarrow {}^{19}K$). These results confirmed earlier findings showing apparent low energy fusion and fission reactions. [2] Moreover, subsequent research with boron indicates apparent low energy fusion reactions in which boron fuses with oxygen to form aluminum and with sulfur to form scandium. [3] At the same time, the QR group has achieved what appear to be low energy transmutations of carbon using carbon-arc under vacuum and in open air. [4] The research group at QR believes these processes can be adapted to accelerate the natural decay cycle of uranium-235, plutonium-239, radium-226, and the fission products cesium-137, iodine-129, and technetium-99, with the long-term potential of reducing the threat posed by radioactive isotopes to human health and the environment.

Uranium-235
Radioactive uranium is the primary constituent of spent nuclear fuel. The half-life of uranium-235 is more than 700 million years. The first step in this process, the alpha decay of uranium-235 into thorium-231 consumes the bulk of this enormous span. The half-lives of the isotopes that follow thorium-231 total approximately 33,000 years with the stable isotope lead-208 as the conclusion of the process.

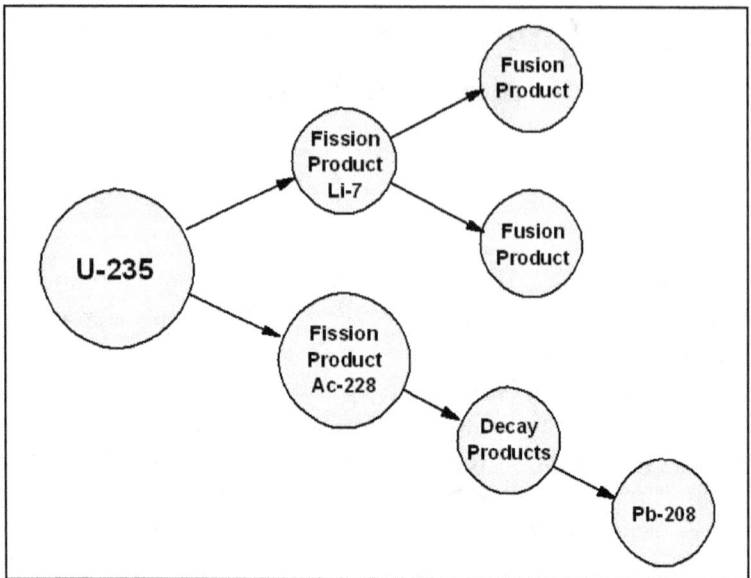
Fig 1. LENR-induced transmutation of uranium-235

The QR research indicates it may be possible to intervene in the decay cycle of uranium in order to reduce the amount of time needed to achieve its transmutation into lead. The most obvious window for intervention is at the beginning of the cycle, by inducing uranium-235 to fission into one of the lighter isotopes in the radioactive decay chain. We propose using lithium, the catalyst element in the studies cited above, as the catalyst for the following low energy fission reaction:

$^{235}U \rightarrow {}^{7}Li + {}^{228}Ac$
Uranium-235 → lithium-7 + actinium-228

According to this hypothesis, the low energy fusion of lithium with sulfur, resulting in potassium, triggers the low energy fission of uranium into lithium and actinium. The low energy fusion reaction can be written as follows:

$^{7}Li + {}^{32}S \rightarrow {}^{39}K$
Lithium-7 + sulfur-32 → potassium-39

These reactions are summarized in Fig. 1. If achieved, they set in motion the natural decay cycle beginning with actinium-228 and ending with lead-208 shown in Fig. 2. Note that the low energy nuclear reaction (LENR) that causes the uranium-235 to fission into actinium-228 results in U-235 being cycled downstream into the natural decay chain of thorium-232. [5]

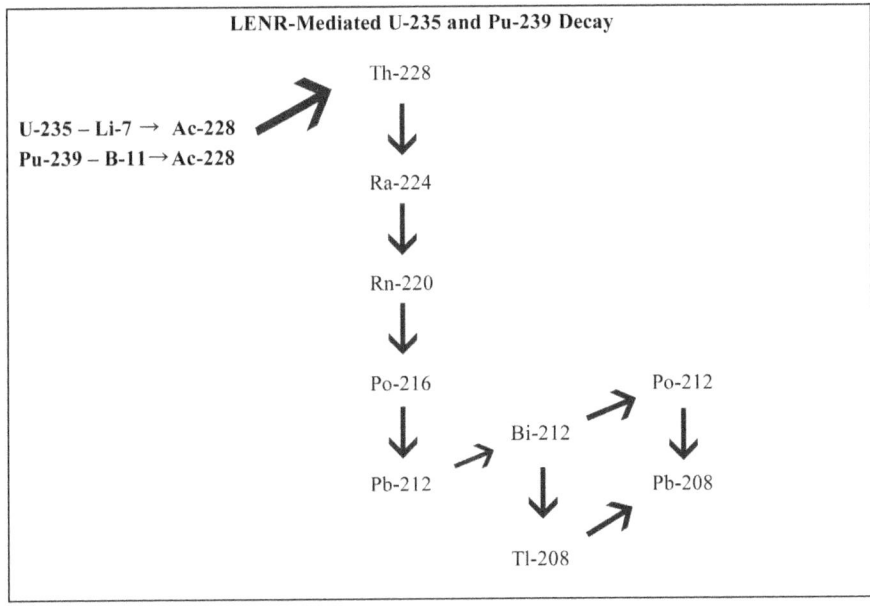

Fig. 2. Accelerated Decay Series: U-235 and Pu-239. Downward arrows represent alpha decay; upward arrows beta decay.

If actinium-228 is produced as predicted, and the natural decay-cycle indicated in Fig. 2 set in motion, the half-life of uranium-235 is compressed from over 700-million years to slightly over 1.9 years. The process is summarized in the formula:

$^{235}U \rightarrow {}^{7}Li + {}^{228}Ac \rightarrow {}^{208}Pb$

Uranium-235 → lithium-7 + actinium-228 (thorium-232 decay cycle) → lead-208

Plutonium-239
There is a significant quantity of deadly plutonium-239 in spent nuclear fuel. Plutonium-239 has a half-life of 24,000 years.

The QR research group achieved promising results with the low energy fusion of boron. A series of experiments for the possible reduction of plutonium-239 similar to the QR boron experiments can be designed using boron as the catalyst element. The low energy fission reaction we propose testing is as follows:

$^{239}Pu \rightarrow {}^{11}B + {}^{228}Ac$
Plutonium-239 → boron-11 + actinium-228

This low energy fission reaction is theoretically triggered by several low energy fusion reactions: [6]

$^{11}B + {}^{16}O \rightarrow {}^{27}Al$
Boron-11 + oxygen-16 → aluminum-27

$^{11}B + {}^{34}S \rightarrow {}^{45}Sc$
Boron-11 + sulfur-34 → scandium-45

Once again, if these LENR were successful in producing actinium-228, like U-235 in the formula described above, Pu-239 would be cycled downstream into the thorium-232 decay chain with the end product being the stable isotope lead-208 (Fig. 2). This process can be summarized as follows:

$^{239}Pu \rightarrow {}^{11}B + {}^{228}Ac \rightarrow {}^{208}Pb$
Plutonium-239 → boron-11 + actinium-228 (thorium-232 decay cycle) → lead-208

Radium-226

Contamination by radium-226 continues to be a problem at U.S. military installations and other sites around the world. Radium-226 is part of the U-238 decay chain with a half-life of 1,600 years. With LENR, it may be possible to compress this time frame considerably by achieving the low energy fission of Ra-226.

QR research on carbon-arc may offer a method for achieving this possibility. Numerous LENR have been reported, both in open air and under vacuum. [7] These low energy fusion reactions could possibly be used to prompt the low energy fission of radium-226, compressing the half-life of radium and accelerating the natural decay cycle from more than 1,600 years to approximately 22 years. (Fig. 3.)

The low energy fission reaction we propose testing is as follows:

$^{226}Ra \rightarrow {}^{12}C + {}^{214}Pb$
Radium-226 → carbon-12 + lead-214

This low energy fission reaction could possibly be triggered by low energy fusion reactions such as those between carbon and oxygen noted in QR carbon-arc research:

$^{12}C + {}^{12}C \rightarrow {}^{24}Mg$
Carbon-12 + carbon-12 → magnesium-24

$^{12}C + {}^{16}O \rightarrow {}^{28}Si$
Carbon-12 + oxygen-16 → silicon-28

$^{12}C + 2({}^{16}O) \rightarrow {}^{44}Ti$
Carbon-12 + 2(oxgyen-16) → titanium-44

$^{12}C + {}^{32}S \rightarrow {}^{44}Ti$
Carbon-12 + sulfur-32 → titanium-44

$2({}^{12}C + {}^{16}O) \rightarrow {}^{56}Fe$ (+2 protons)
2(Carbon-12 + oxygen-16) → iron-56 + two protons

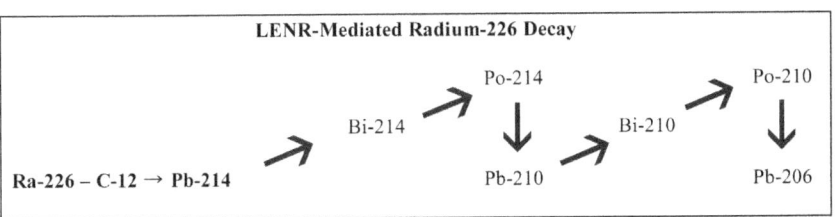

Fig. 3. Accelerated Decay Series: Ra-226. Downward arrows represent alpha decay; upward arrows beta decay.

Cesium-137

Cesium-137, a product of nuclear fission is a major radionuclide in spent nuclear fuel. It is of major concern for Department of Energy environmental management sites and has a half-life of 30 years. It decays by emitting a beta particle.

It's decay product, barium-137m (the "m" is for metastable) stabilizes by emitting an energetic gamma ray with a half-life of approximately 2.6 minutes. It is this decay product that qualifies cesium-137 as a radiation hazard.

The environmental dangers posed by cesium-137 were highlighted by the crisis at Fukushima Daichi reactor in Japan. Writing in the Proceedings of the National Academy of Sciences, [8] an international team of scientists described the threat posed by cesium-137:

"The largest concern on the cesium-137 (^{137}Cs) deposition and its soil contamination due to the emission from the Fukushima Daiichi Nuclear Power Plant (NPP) showed up after a massive quake on March 11, 2011. Cesium-137 (^{137}Cs) with a half-life of 30.1 y causes the largest concerns because of its deleterious effect on agriculture and stock farming, and, thus, human life for decades. Removal of ^{137}Cs contaminated soils or land use limitations in areas where removal is not possible is, therefore, an urgent issue."

Contamination by cesium-137 was a major problem following the Chernobyl disaster. As John Emsley states: [9]

"Uranium fuel rods in nuclear power stations produce cesium-137. The half-life of cesium-137 is 30 years, which means that it takes over 200 years to reduce it to 1% of its former level. For this reason, an accident at a nuclear power plant can contaminate the environment around for generations, which is why the Chernobyl accident in the Ukraine in 1986 was such an environmental disaster. It released a large amount of radioactive cesium-137 which drifted all over Western Europe, affecting sheep farms as far west as Scotland, Ireland, and Wales, over 1500 miles from the accident. There it was washed to earth by heavy rain and taken up by the roots of plants, thus becoming part of the vegetation that sheep ate."

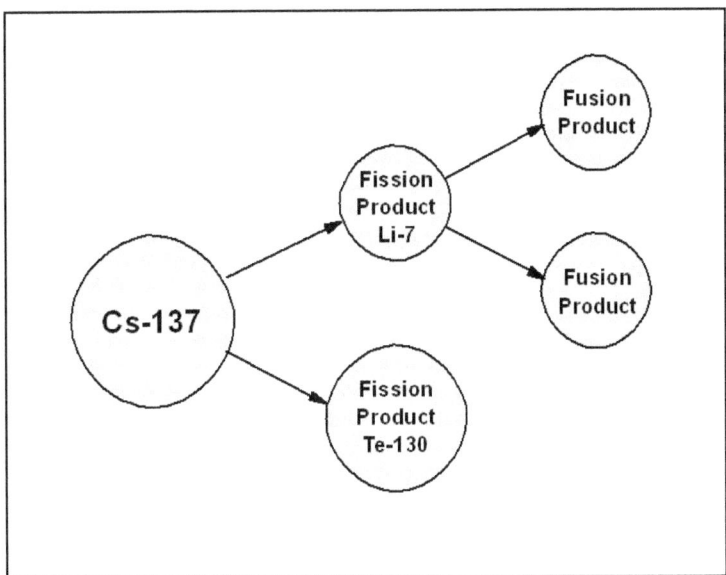
Fig. 4. LENR-induced transmutation of cesium-137

Using LENR, it may be possible to convert cesium-137 to tellurium-130, a stable non-radioactive isotope, thus redirecting and compressing the cesium-137 decay cycle (Fig. 4). The LENR-induced fission formula is as follows:

$^{137}Cs \rightarrow {}^7Li + {}^{130}Te$
Cesium-137 → lithium-7 + tellurium-130

In theory, the low energy fission reaction would be triggered by the low energy fusion of lithium and sulfur:

$^7Li + {}^{32}S \rightarrow {}^{39}K$
Lithium-7 + sulfur-32 → potassium-39

In a separate experiment, cesium-137 may also transmute into neodymium-148 through a low energy fusion reaction:

$^{137}Cs + {}^{11}B \rightarrow {}^{148}Nd$
Cesium-137 + boron-11 → neodymium-148

If the fusion reaction can be proven and scaled to production levels, it would then be possible to convert dangerous radioactive waste into a valuable rare earth metal widely utilized today in the magnets in hybrid vehicles.

Iodine-129
Iodine-129 is a long-lived isotope of iodine created primarily from the fission of uranium and plutonium in nuclear reactors. It decays with a half-life of 15.7 million years. Significant amounts were released into the atmosphere following nuclear weapons tests in the 1950s and 1960s. Iodine-129 is long-lived and mobile in the environment and is thus of special importance in disposal and management of spent nuclear fuel.

It may be possible to compress the natural decay cycle of this radio-isotope through the process of low energy fission. The LENR-induced fission reaction is as follows:

$^{129}I \rightarrow {}^7Li + {}^{122}Sn$
Iodine-129 → lithium-7 + tin-122

Once again, according to theory, low energy fusion of lithium and sulfur would serve as the catalyst for the reaction:

$^7Li + {}^{32}S \rightarrow {}^{39}K$
Lithium-7 + sulfur-32 → potassium-39

During the experiment, iodine-129 may also transmute into barium-146 through a simultaneous fission reaction:

$^{129}I + {}^7Li \rightarrow {}^{136}Ba$
Iodine-129 + lithium-7 → barium-136

Technetium-99
Technetium-99 is radioisotope of technetium that decays with a half-life of 211,000 years to stable ruthenium-99. It is the most significant long-lived fission product of uranium-235.

Its high fission yield, relatively long half-life, and mobility in the environment make technetium-99 one of the more problematic components of nuclear waste. There have been releases into the environment from atmospheric nuclear tests, nuclear reactors, and in the late 1990s from the Sellafield plant, which released nearly 1,000 kg into the Irish Sea.

It may be possible to accelerate the half–life of Tc-99 by inducing the following low energy fission reaction:

$^{99}Tc \rightarrow {}^{7}Li + {}^{92}Zr$
Technetium-99 → lithium-7 + zirconium-92

Once again, in theory, the reaction would be triggered by the low energy fusion of lithium and sulfur:

$^{7}Li + {}^{32}S \rightarrow {}^{39}K$
Lithium-7 + sulfur-32 → potassium-39

During the experiment, Tc-99 may also transmute into Pd-106 through the following fusion reaction:

$^{99}Tc + {}^{7}Li \rightarrow {}^{106}Pd$
Technetium-99 + lithium-7 → palladium-106

Guidelines for Methodology
The experiments on low energy transmutation cited in the Abstract can serve as a starting point for designing experiments to test the nuclear reduction hypothesis presented in this paper. [10] A vacuum tube similar to that used in the QR low energy transmutation tests and shown in Fig. 5 can be considered for the nuclear reduction tests. Because silver is a strong conductor of electricity and a neutron absorber, we propose using it as the anode and cathode material, with other test materials adjusted for each experiment as indicated below (Fig. 6). Moreover, silver may react independently with lithium to form tin ($^{109}Ag + {}^{7}Li \rightarrow {}^{116}Sn$). This reaction was noted in a previous QR test. [11] Keep in mind that these suggestions are guidelines only, based on previous low energy fusion and fission experiments. They will need to be adjusted in real time based upon further study and experience.

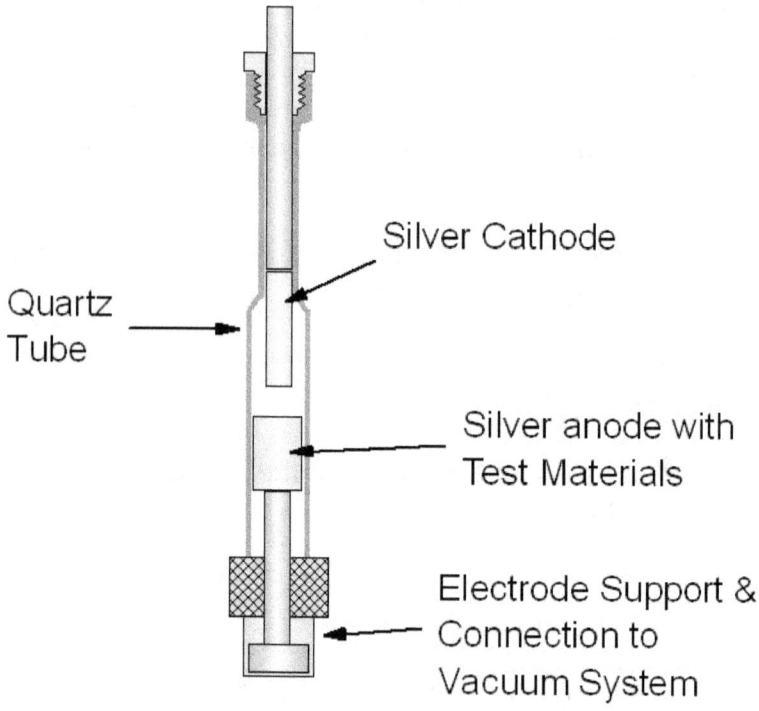

Fig. 5. Tube and electrode configuration

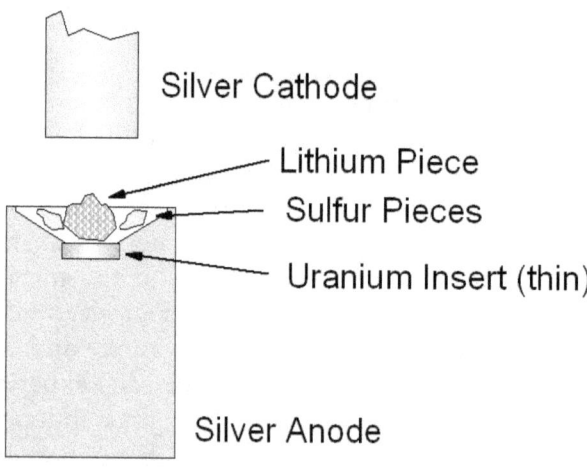

Fig. 6. Electrodes and test material suggested for the U-235 → Li-7 + Ac-228 experiment

Uraninum-235:

$^{235}U \rightarrow {}^{7}Li + {}^{228}Ac \rightarrow {}^{208}Pb$

Electrodes made of Ag
Test Materials:

1. Uranium insert (thin wafer or foil) in anode
2. Lithium test material
3. Sulfur test material
4. Pure neon/oxygen backfill

Plutonium-239:

$^{239}Pu \rightarrow {}^{11}B + {}^{228}Ac \rightarrow {}^{208}Pb$

Electrodes made of Ag
Test Materials:

1. Plutonium insert (thin wafer or foil) in anode
2. Boron test material
3. Sulfur test material (optional)
4. Pure neon/oxygen backfill

Radium-226:

$^{226}Ra \rightarrow {}^{12}C \rightarrow {}^{214}Pb \rightarrow {}^{206}Pb$

Electrodes made of Ag
Test Materials:

1. Radium insert (thin wafer or foil) in anode
2. Carbon (graphite) test material
3. Sulfur test material
4. Pure nitrogen/oxygen backfill*

*Note: Adding nitrogen allows the process to take advantage of potential carbon-nitrogen reactions such as those noted in QR research. [12]

Cesium-137:

$^{137}Cs \rightarrow {}^7Li + {}^{130}Te$

Electrodes made of Ag
Test Materials:

1. Cesium insert (thin wafer or foil) in anode
2. Lithium test material
3. Sulfur test material
4. Pure neon/oxygen backfill

$^{137}Cs + {}^{11}B \rightarrow {}^{148}Nd$

Electrodes made of Ag
Test Materials:

1. Cesium insert (thin wafer or foil) in anode
2. Boron test material
3. Sulfur test material
4. Pure neon/oxygen backfill

Iodine-129:

$^{129}I \rightarrow {}^7Li + {}^{122}Sn$
$^{129}I + {}^7Li \rightarrow {}^{136}Ba$

Electrodes made of Ag
Test Materials:

1. Iodine inserted in or on anode
2. Lithium test material
3. Sulfur test material
4. Pure neon/oxygen backfill

Technetium-99:

$^{99}Tc \rightarrow {}^7Li + {}^{92}Zr$
$^{99}Tc + {}^7Li \rightarrow {}^{106}Pd$

Electrodes made of Ag
Test Materials:

1. Technetium insert (thin wafer or foil) in anode
2. Lithium test material
3. Sulfur test material
4. Pure neon/oxygen backfill

Procedure for the Above Experiments:

1. Insert is placed on or into the anode.
2. Measured quantity of test materials are placed in anode recess.
3. Glass/quartz tube is placed over the anode assembly.
4. Cathode is inserted into the tube and secured at the desired separation from the anode.
5. Fill with neon (or nitrogen for Ra-226) to 2 torr.
6. Strike plasma using direct current (DC).
7. Admit oxygen fill to 6 torr. Continue until reaction noticeably slows or tube is in danger of breaking (approximately 10-20 minutes.)
8. Disconnect power and allow sample to cool.

Conclusion

As of this writing, the problem of nuclear waste disposal remains unsolved. In an op-ed published in the *Santa Monica Daily Press*, [13] Dr. Jeffrey Patterson, former head of Physicians for Social Responsibility (PSR) stated:

"2011 was a scary year for nuclear reactor sites. The summer floods threatened to encroach on reactors in Nebraska and Iowa, an earthquake and a hurricane happened in quick succession to rattle and flood the East Coast, and the continuing events of the Fukushima-Daichi reactor accident provided harrowing examples of the threats posed to spent fuel at reactor sites. The fate of spent fuel there kept the world on edge for days. It's worth noting that the amount of fuel in vulnerable storage pools in Japan was far less than what is crowded into pools at many U.S. reactors. As we all learned, a loss of coolant could produce a fuel melt and large radiation release.

"It wasn't supposed to be this way. Used reactor fuel was to be permanently stored in deep underground repositories, away from floods and other natural hazards. But the solution to the nation's nuclear waste problem has been elusive for decades. Meanwhile, 65,000 metric tons of spent reactor fuel is still looking for a home."

The Blue Ribbon Commission on America's Nuclear Future proposes transferring spent nuclear fuel, now scattered at 70 locations around the U.S., to temporary storage areas, pending selection of more permanent deep geologic repositories. This proposal is not without controversy. As Dr. Patterson states:

"Moving spent fuel around the country is not a risk worth taking. Rather than addressing the problem, an "interim" facility would only relocate it. So what is the best option? Hardened on-site storage of spent fuel. It's safe, cost-effective — and readily available. PSR and over 170 public interest organizations from all 50 states are calling for adoption of this approach. Storing reactor fuel at reactor sites in hardened buildings that can resist severe attacks, such as a direct hit by high-powered explosives or a large aircraft, as is done in Germany, offers the safest and most sensible option until a permanent repository can be found."

These proposals offer opportunities for research on LENR-induced transmutation. Research laboratories could be set up at future on-site hardened facilities or even now at current waste storage sites, as well as at future interim facilities. These laboratories can begin first-round investigation of LENR-induced transmutation. If successful, scale-up can proceed to levels required to reduce the on-site, regional, and global inventory of nuclear waste.

Moreover, LENR-induced transmutation may offer an efficient low-cost alternative to accelerator transmutation of waste (ATW). In 1999, the U.S. Department of Energy's (DOE) Office of Civilian Radioactive Waste Management submitted a report to Congress entitled "A Roadmap for Developing Accelerator Transmutation of Waste (ATW) Technology." Sekazi K. Mtingwa of MIT describes this approach as follows: [14]

"Transmutation means the transformation of one atom into another by changing its nuclear structure. In the present context this means bombarding a highly radioactive atom with neutrons, preferably fast neutrons, from either a fast nuclear reactor or spallation neutrons created by bombarding protons from a high-energy accelerator on a suitable target."

The Oak Ridge National Laboratory (ORNL) is currently investigating methods for accelerator transmutation (ATW) of nuclear wastes. An article in the ORNL *Review* states: [15]

"Conceived by scientists at Los Alamos National Laboratory, ATW uses a linear accelerator system to produce neutrons for transmutation of excess weapons plutonium and other radioactive DOE wastes, such as technetium-99 and iodine-129.

"Ultimately, the potential of partitioning and transmutation to waste management is this: If a radioactive waste stream no longer exists, then it poses no radiological hazard. More than anything else, this simple fact has spurred the recent resurgence of interest in partitioning-transmutation technology."

Meanwhile, in the Eurozone, the European nuclear establishment is pressing ahead with a $1.2 billion R&D project to look into high-energy neutron-induced transmutation. The first stage of the project, the setup of a demonstration system known as "Guinevere" that combines a particle accelerator and a nuclear reactor, took place in January 2012 at the Belgian Nuclear Research Center at Mol. A larger version of the reactor system, known as Myrrha (Multipurpose Hybrid Research Reactor for High-tech Applications), is scheduled to bcome operational in 2023. A press release from the World Nuclear Association explains the thinking behind the project: [16]

"Myrrha will be able to produce radioisotopes and doped silicon, but its research functions would be particularly well suited to investigating transmutation. This is when certain radioactive isotopes with long half-lives are made to 'catch' a neutron and thereby change into a different isotope that will decay more quickly to a stable form with no radioactivity."

"If achievable on an industrial scale, transmutation could greatly simplify the permanent geologic disposal of radioactive waste."

The Quantum Rabbit group estimates that research on LENR-induced transmutation could begin at a fraction of the estimated $1.2 billion startup cost of the Myrrha project. (QR estimates $1.2 million for feasibility study and $12 million to develop a prototype system, amounts that are respectively 0.1% and 1% the cost of Myrrha.) Rather than a highly centralized billion-dollar processing system, LENR-induced transmutation technology could be distributed to nuclear power stations around the globe at an affordable cost. The task of nuclear remediation would become the responsibility of the individual power station and thus remain local instead of becoming highly centralized. Also, the amount of power needed to conduct LENR-induced transmutation would be miniscule compared to the power required to operate a particle accelerator and nuclear reactor. At the very least, research on LENR-induced transmutation should proceed on a parallel track to the high-energy neutron-induced transmutation projects currently underway or under consideration in order to determine which approach yields the most promising results.

[1] Esko, Edward, "Anomalous Metals Part II," *Infinite Energy*, No. 103, 2012.
[2] Esko, Edward and Jack, Alex, *Cool Fusion*, second edition, Amber Waves, Becket, Mass., USA, pp. 56-113, 132-151, 2012.
[3] Esko, Edward, "In Search of the Platinum Group Metals Part II," *Infinite Energy*, No. 104, 2012.
[4] Esko, Edward and Jack, Alex, *Cool Fusion*, second edition, Amber Waves, Becket, Mass., USA, pp. 56-60, 88-97, 2012.
[5] Argonne National Laboratory, "Human Health Fact Sheet," Fig. N.3 Natural Decay Series: Thorium-232, 2005.
[6] Esko, Edward, "In Search of the Platinum Group Metals Part II," *Infinite Energy*, no. 104, 2012.
[7] Esko, Edward and Jack, Alex, *Cool Fusion*, second edition, Amber Waves, Becket, Mass., USA, pp. 56-60, 88-97, 2012.
[8] Yasunari, Teppei J., Stohl, Andreas, Hayano, Ryugo S., Burkhart, John F., Eckhardt, Sabine, Yasunari, Tetsuzo, "Cesium-137 deposition and contamination of Japanese soils due to the Fukushima nuclear accident," Proceedings of the National Academy of Sciences, November, 14, 2011.
[9] Emsley, John, *Nature's Building Blocks: An A-Z Guide to the Elements*, first edition, Oxford University Press, Oxford, England, pp. 82, 2001.
[10] Refer to Occupational Safety and Health Administration (OSHA) guidelines for the handling of hazardous materials prior to initiating these experiments.

[11] Esko, Edward and Jack, Alex, *Cool Fusion*, second edition, Amber Waves, Becket, Mass., USA, pp. 56-113, 132-151, 2012.
[12] Esko, Edward and Jack, Alex, *Cool Fusion*, second edition, Amber Waves, Becket, Mass., USA, pp. 79-87, 2012.
[13] Patterson, Jeffrey, "Time to fix our nuclear waste disposal system," *Santa Monica Daily Press*, January 6, 2012.
[14] Mtingwa, Sekazi K., "Feasibility of Transmutation of Radioactive Isotopes," An International Spent Nuclear Fuel Storage Facility -- Exploring a Russian Site as a Prototype: Proceedings of an International Workshop, The National Academies Press, Washington, D.C., 2005.
[15] Michaels, Gordon E., "Partitioning and Transmutation: Making Wastes Nonradioactive," Oak Ridge National Laboratory Review, Vol. 44, No. 2, 2011.
[16] World Nuclear News, World Nuclear Association, January 11, 2012.

Source: Edward Esko, "LENR-Induced Transmutation of Nuclear Waste," *Infinite Energy*, No. 104, 2012.

Ten Transmutation Experiments

Abstract
In ten vacuum arc experiments conducted over a three-and a half-year period, investigators at Quantum Rabbit LLC (QR) noted evidence for possible low energy transmutation, including the consistent and anomalous appearance of gold, germanium, and potassium. All ten experiments employed the same inputs and utilized two vacuum tube designs, one horizontal and the other vertical. Papers on experiments conducted on July 30, 2009 and September 27, 2011 have been published in *Infinite Energy* ("Anomalous Metals in Electrified Vacuum," No. 99, 2011, and "Anomalous Metals Part II," No. 103, 2012.) A paper on the September 27, 2011 experiment was also presented at ICCF-17 (International Conference on Cold Fusion) held in South Korea in August 2012. Note that the numbers assigned to an experiment in this paper are not necessarily those assigned to the experiment during the lab session in which it was performed.

Horizontal Tube Experiments

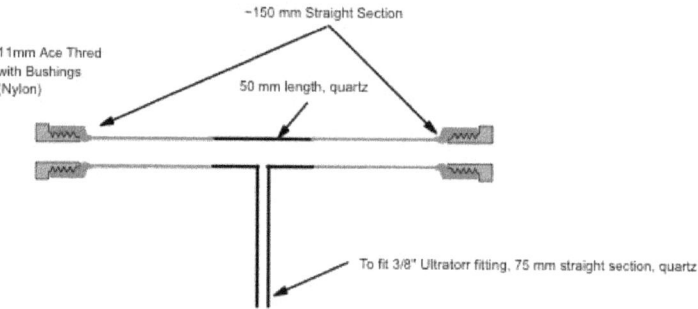

Fig. 1. The horizontal tube design used in Tests 1 – 4. A borosilicate glass tube 150 mm long with a 50 mm quartz midsection. A 3/8-inch diameter quartz straight section, perpendicular to the tube and 75 mm in length, connects the midsection with the vacuum manifold and serves as the entrance for fill gas.

Test 1
Location: QR Lab, Nashua, NH
Date: December 30, 2008

Tube: Horizontal Design (Fig. 1)
Electrodes: Copper (Cu)
Electrode Insert: Lead (Pb) in Anode
Test Materials:
    1. Lithium (Li)
    2. Sulfur (S)
Fill Gas: Oxygen (O$_2$)
ICP Analysis: NHML File No. 25985 (Table 1)
Anomalies: Au – 5540 ppm; Ge – 416 ppm; K – 341 ppm

Procedure: Copper electrodes were cut from a high purity copper rod from Alfa Aesar. The Puratronic® rod was composed of 99.999% pure copper (metals basis.) A high purity lead slug (99.9999% metals basis) was used to create a lead insert pressed into a 0.27-inch diameter cavity in the center of the copper anode.

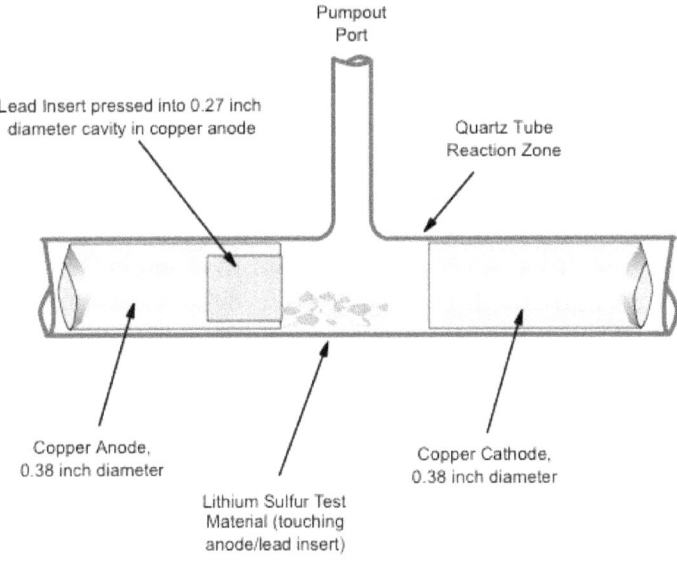

Fig. 2. Tube and electrode configuration used in Tests 1–4

The copper cathode was inserted in one end of the tube. Sulfur granules were placed in the quartz midsection, together with pieces of lithium cut from a lithium rod. The sulfur, listed in the Alfa Aesar catalogue as "sulfur pieces," was actually a coarse yellow powder that was difficult to maneuver into the tube.

However, after persistent attempts, we succeeded in placing a sufficient quantity in the tube. Like the copper electrodes, the sulfur was labeled Puratronic® and certified by Alfa Aesar as being 99.999% pure (metals basis.) The lithium pieces were also from Alfa Aesar and certified as being 99.9% pure (metals basis).

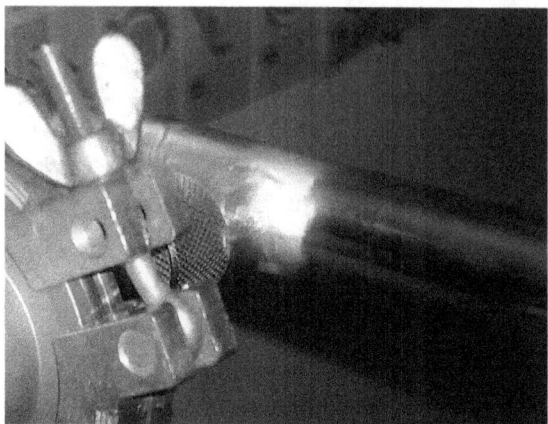

Fig. 3. The glow discharge in Test 1

After placing the sulfur and lithium in the tube, the lead-tipped anode was inserted at the opposite end, and the electrodes connected to the power supply. The lead-tipped anode was in contact with the lithium/sulfur powder (Fig. 2). The tube was pumped down to vacuum and backfilled with oxygen to approximately 3.5 torr. Direct current was passed through the electrodes, producing an electric arc (approximately 4 amps) and glow discharge (Fig. 3). After about 12 minutes, the current was turned off, the vacuum pumps disconnected, and the sample allowed to cool.

Independent Analysis: Three samples were retrieved: the lead-tipped anode, the copper cathode, and sulfur-lithium residue from the center of the tube. The anode and cathode tips had obviously undergone changes during the experiment, with the presence of an undetermined residue on their surface.

Samples were labeled, packaged, and sent to New Hampshire Materials Laboratory with instructions to extract material for analysis by scraping the surfaces of the lead anode and copper cathode, and collecting the residue at the center of the tube.

Results of analysis by Inductively Coupled Plasma Atomic Spectroscopy (ICP) are shown in Table 1.

Table 1. NHML Test Report for Test 1 January 9, 2009

Test 1	Lead Anode	Copper Cathode	S-Li Residue
Gold	5300 ppm	130 ppm	110 ppm
Germanium	260 ppm	50 ppm	106 ppm
Potassium	230 ppm	73 ppm	38 ppm

Tests 2-4
Location: QR Lab, Owls Head, ME
Date: July 30, 2009
Tube: Horizontal Design (Fig. 1)
Electrodes: Copper (Cu)
Electrode Insert: Lead (Pb) in Anode
Test Materials:
   1. Lithium (Li)
   2. Sulfur (S)
Fill Gas: Oxygen (O^2)
ICP Analysis: NHML File 26657 (Table 2)
Anomalies: Au – 174 ppm; Ge – 3596 ppm; K – 750 ppm

Procedure: All three experiments were essentially duplicates of Test 1. Copper electrodes were cut from a high purity copper rod from Alfa Aesar. The Puratronic® rod was composed of 99.999% pure copper (metals basis.) A high purity lead slug (99.9999% metals basis) was used to create a lead insert pressed into a 0.27-inch diameter cavity in the center of the copper anode.

The copper cathode was inserted in one end of the tube. Sulfur granules were placed in the quartz midsection, together with pieces of lithium cut from a lithium rod. Like the copper electrodes, the sulfur was labeled Puratronic® and certified by Alfa Aesar as being 99.999% pure (metals basis.) The lithium pieces were also from Alfa Aesar and certified as being 99.9% pure (metals basis).

After placing the sulfur and lithium in the tube, the lead-tipped anode was inserted at the opposite end, and the electrodes connected to the power supply. The lead-tipped anode was in contact with the lithium/sulfur powder (Fig. 2.) The tube was pumped down to vacuum and backfilled with oxygen to approximately 3.5 torr.

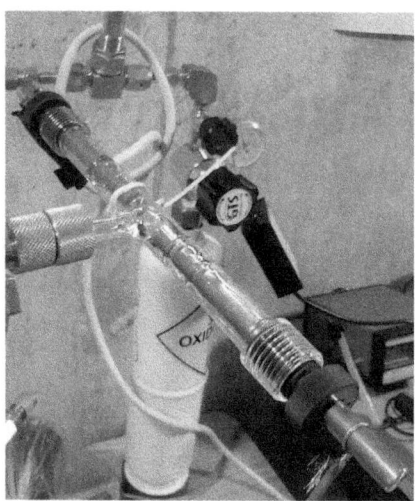

Fig. 4. Equipment for Tests 2–4

Direct current was passed through the electrodes, producing an electric arc (approximately 4 amps) and glow discharge. After about 12 minutes, the current was turned off, the vacuum pumps disconnected, and the sample allowed to cool.

Table 2. NHML Test Report for Tests 2–4, August 14, 2009

	Lead Anode	Copper Cathode	S-Li Residue
Test 2			
Gold	162 ppm	11 ppm	1 ppm
Germanium	1018 ppm	2190 ppm	388 ppm
Potassium	638 ppm	85 ppm	27 ppm
Test 3			
Gold	5 ppm	<1 ppm	<1 ppm
Germanium	102 ppm	119 ppm	31 ppm
Potassium	16 ppm	31 ppm	6 ppm
Test 4			
Gold	7 ppm	<1 ppm	<1 ppm
Germanium	35 ppm	2 ppm	7 ppm
Potassium	17 ppm	10 ppm	1 ppm

Independent Analysis: Three samples were retrieved from each test: the lead-tipped anode, the copper cathode, and sulfur-lithium residue from the center of the tube. The anode and cathode tips had obviously undergone changes during the experiment, with the presence of an undetermined residue on their surface. The three sets of samples were labeled, packaged, and sent to New Hampshire Materials Laboratory with instructions to extract material for analysis by scraping the surface of the lead anode and copper cathode, and collecting the residue at the center of the tube. Results of analysis by Inductively Coupled Plasma Atomic Spectroscopy (ICP) are shown in Table 2.

Vertical Tube Experiments

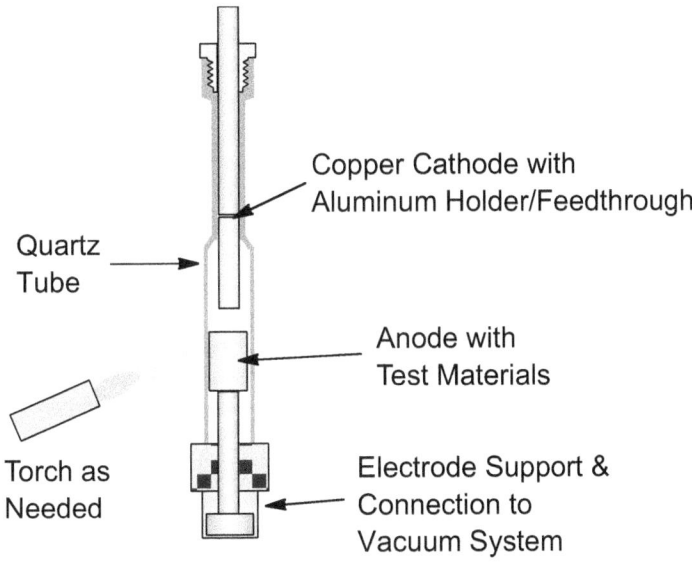

Fig. 4. The vertical tube design used in Tests 5–10. Shown are the copper cathode assembly, copper anode, quartz tube, and vacuum system connection. The torch was not used in these tests.

Test 5
Location: QR Lab, Owls Head, ME
Date: September 27, 2011
Tube: Vertical Design (Fig. 4)
Electrodes: Copper (Cu)
Electrode Insert: Lead (Pb) in Anode

Test Materials:
   1. Lithium (Li)
   2. Sulfur (S)
Fill Gas: Neon (Ne) and Oxygen (O_2)
ICP Analysis: NHML File 28929 (Table 3)
Anomalies: Au – 252 ppm; K – 181 ppm

Procedure: Test 5 employed the same inputs as Tests 1-4, with several new features. The first modification was the creation of a small recess at the center of the anode. The recess facilitated a more secure placement of test material in the tube. The recess helped confine the test material to the reaction zone. As was the case in 2009, a lead insert was placed in the center of the copper anode. The lead insert consisted of a lead slug approximately 0.25-inch diameter pressed into a 0.25-inch by 0.25-inch drilled hole. One piece of lithium was centrally placed atop the lead insert. The lithium was surrounded by sulfur pieces (Fig. 5 and Fig. 6). The second modification was the use of neon as a fill gas to strike plasma before admitting oxygen as the catalyst. Oxygen was the sole fill gas in the first four tests.

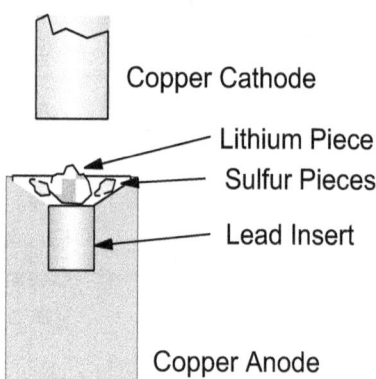

Test 1, Sept 27, 2011

Fig. 5. Electrodes, insert, and test material used in Tests 5-10

The test proceeded in real time as follows (keep in mind that the data points are approximate.) The tube was pumped down to vacuum. Neon was admitted at the start to 2 torr. At 1 minute in, the torr reading was 3.0, while power supply readings measured 53 volts and 6.95 amps. The inside of the tube was glowing red-purple, with what appeared to be the color of neon plasma. At about 2 minutes, the readings were as follows: 3.0 torr, 70 volts, and 5.63 amps. Intense heat was generated at this point; so that the test materials appeared to be melting.

Fig. 6. Lithium and sulfur test material atop copper anode in Test 5

Oxygen was admitted between minute 3 and 4, and the torr reading went up to approximately 8.28. Following the oxygen fill, the test material began glowing a ruby red, the characteristic color of lithium plasma. At around 6 minutes, there was concern that the tube had failed. Power was disconnected. Thirty seconds later it was decided that the tube was still viable, and the decision was made to admit fresh oxygen and fire up the tube once again. At this point the tube began glowing blue-green. Between 8 to 9 minutes, the power alternated between 45-55 volts and 6.65 and 7.5 amps.

After 10 minutes, conditions in the tube appeared to stabilize. Voltage hovered around 77 and amps around 5.4. The test finished after a total time of approximately 14 minutes with a 3.5 torr reading. Upon conclusion of the experiment, the electricity was turned off, the vacuum pumps disconnected, and the samples allowed to cool.

Independent Analysis: Samples retrieved for testing included the lead-tipped copper anode with lithium-sulfur residue in its center, the copper cathode, and the quartz tube itself, which contained residue on its inner surface. The anode, cathode, and inside of the tube had undergone noticeable changes during the experiment. Samples were labeled, packaged, and sent to New Hampshire Materials Laboratory with instructions to extract material for analysis by 1) scraping the surface of the lead anode; and 2) scraping residue from the copper cathode and inner surface of the tube. Results of analysis by Inductively Coupled Plasma Atomic Spectroscopy (ICP) are shown in Table 3.

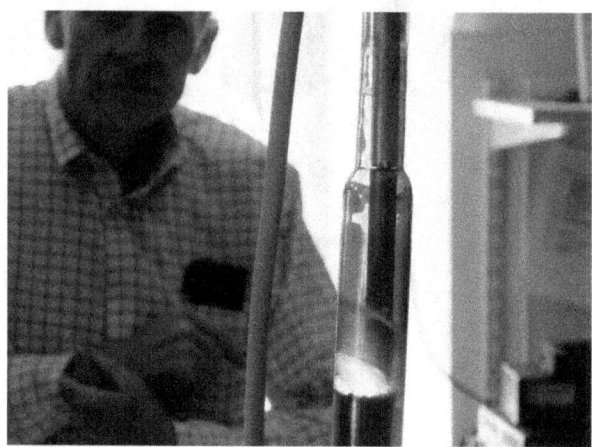

Fig. 7. The glow discharge in Test 5

Table 3. NHML Test Report for Test 5 October 14, 2011

Test 5	Anode/Residue	Cathode/Tube
Gold	252 ppm	<0.01 ppm
Potassium	181 ppm	<0.01 ppm
Germanium	<0.01 ppm	<0.01 ppm

Tests 6-10

Location: QR Lab, Owls Head, ME
Date: April 11, 2012
Tube: Vertical Design (Fig. 5)
Electrodes: Copper (Cu)
Electrode Insert: Lead (Pb) in Anode

Test Materials:
  1. Lithium (Li)
  2. Sulfur (S)

Fill Gas: Neon (Ne) and Oxygen (O_2)
Analysis: ICP NHML File 29508 (Table 4)
Anomalies: K – 24.37 ppm

Procedure: Tests 6–10 were essentially repeat tests of Test 5 utilizing the vertical tube design. Placement of the copper electrodes and lithium and sulfur test materials were the same, as was the use of neon and oxygen fill, power delivered to the tube, and level of vacuum attained. These factors varied slightly in each experiment. The protocol used for all five experiments was as follows:

1. Lithium/sulfur test material mixed and inserted in anode recess.
2. Pump down to vacuum.
3. Fill with neon to 2 torr.
4. Strike plasma and stabilize at approx. 3 torr.
5. After 4 minutes admit oxygen fill to 6 torr.
6. Continue until reaction noticeably slows or tube is in danger of breaking.
7. Disconnect power after approx. 15 minutes and allow sample to cool.

Fig. 8. Steve Hansen admits fill gas to the vertical tube in Test 7

Independent Analysis: Samples retrieved for analysis from each of the five experiments included the lead-tipped copper anode with lithium-sulfur residue in its center, the copper cathode, and the quartz tube itself, which contained residue on its inner surface.

The samples were labeled, packaged, and sent to New Hampshire Materials Laboratory with instructions to extract material for analysis by 1) scraping the surface of the lead anode; and 2) scraping residue from the copper cathode and inner surface of the tube. Results of analysis by Inductively Coupled Plasma Atomic Spectroscopy (ICP) are shown in Table 4.

Table 4. NHML Test Report for Tests 6–10, May 4, 2012

Test 6	Anode/Residue	Cathode/Tube
Gold	<0.01 ppm	<0.01 ppm
Potassium	6.09 ppm	<0.01 ppm
Germanium	<0.01 ppm	<0.01 ppm
Test 7	Anode/Residue	Cathode/Tube
Gold	<0.01 ppm	<0.01 ppm
Potassium	6.50 ppm	<0.01 ppm
Germanium	<0.01 ppm	<0.01 ppm
Test 8	Anode/Residue	Cathode/Tube
Gold	<0.01 ppm	<0.01 ppm
Potassium	24.37 ppm	<0.01 ppm
Germanium	<0.01 ppm	<0.01 ppm
Test 9	Anode/Residue	Cathode/Tube
Gold	<0.01 ppm	<0.01 ppm
Potassium	11.66 ppm	<0.01 ppm
Germanium	<0.01 ppm	<0.01 ppm
Test 10	Anode/Residue	Cathode/Tube
Gold	<0.01 ppm	<0.01 ppm
Potassium	4.41 ppm	<0.01 ppm
Germanium	<0.01 ppm	<0.01 ppm

Interpretation

In *Infinite Energy* (No. 106) David Nagel describes the type of transmutation experiments presented above:

"LENR experiments have produced two major types of transmutation data. The first and most common category requires determination of the absolute or relative amounts of specific elements or isotopes by pre-and post-run analysis of elemental or isotopic concentrations, ideally with spatial resolution.

Test Materials:
 1. Lithium (Li)
 2. Sulfur (S)
Fill Gas: Neon (Ne) and Oxygen (O_2)
Analysis: ICP NHML File 29508 (Table 4)
Anomalies: K – 24.37 ppm

Procedure: Tests 6–10 were essentially repeat tests of Test 5 utilizing the vertical tube design. Placement of the copper electrodes and lithium and sulfur test materials were the same, as was the use of neon and oxygen fill, power delivered to the tube, and level of vacuum attained. These factors varied slightly in each experiment. The protocol used for all five experiments was as follows:

1. Lithium/sulfur test material mixed and inserted in anode recess.
2. Pump down to vacuum.
3. Fill with neon to 2 torr.
4. Strike plasma and stabilize at approx. 3 torr.
5. After 4 minutes admit oxygen fill to 6 torr.
6. Continue until reaction noticeably slows or tube is in danger of breaking.
7. Disconnect power after approx. 15 minutes and allow sample to cool.

Fig. 8. Steve Hansen admits fill gas to the vertical tube in Test 7

Independent Analysis: Samples retrieved for analysis from each of the five experiments included the lead-tipped copper anode with lithium-sulfur residue in its center, the copper cathode, and the quartz tube itself, which contained residue on its inner surface.

The samples were labeled, packaged, and sent to New Hampshire Materials Laboratory with instructions to extract material for analysis by 1) scraping the surface of the lead anode; and 2) scraping residue from the copper cathode and inner surface of the tube. Results of analysis by Inductively Coupled Plasma Atomic Spectroscopy (ICP) are shown in Table 4.

Table 4. NHML Test Report for Tests 6–10, May 4, 2012

Test 6	Anode/Residue	Cathode/Tube
Gold	<0.01 ppm	<0.01 ppm
Potassium	6.09 ppm	<0.01 ppm
Germanium	<0.01 ppm	<0.01 ppm
Test 7	Anode/Residue	Cathode/Tube
Gold	<0.01 ppm	<0.01 ppm
Potassium	6.50 ppm	<0.01 ppm
Germanium	<0.01 ppm	<0.01 ppm
Test 8	Anode/Residue	Cathode/Tube
Gold	<0.01 ppm	<0.01 ppm
Potassium	24.37 ppm	<0.01 ppm
Germanium	<0.01 ppm	<0.01 ppm
Test 9	Anode/Residue	Cathode/Tube
Gold	<0.01 ppm	<0.01 ppm
Potassium	11.66 ppm	<0.01 ppm
Germanium	<0.01 ppm	<0.01 ppm
Test 10	Anode/Residue	Cathode/Tube
Gold	<0.01 ppm	<0.01 ppm
Potassium	4.41 ppm	<0.01 ppm
Germanium	<0.01 ppm	<0.01 ppm

Interpretation

In *Infinite Energy* (No. 106) David Nagel describes the type of transmutation experiments presented above:

"LENR experiments have produced two major types of transmutation data. The first and most common category requires determination of the absolute or relative amounts of specific elements or isotopes by pre-and post-run analysis of elemental or isotopic concentrations, ideally with spatial resolution.

"Observations that the quantities of elements after an experiment are greater than the amounts present before the experiment are taken as evidence of nuclear reactions. This class of LENR experiment requires careful and often expensive low-level analyses of several elements or isotopes of interest before and after an experiment. Besides the labor and cost of such work, there is an enduring problem that the apparent increase in a concentration might be due to production of non-uniform distribution of the entity of interest during the experiment. That is, no new atoms might be produced, but atoms already present and not detected before the experiment can be brought above detection thresholds, or into a region subjected to analysis, by processes during the experiment."

Let us examine the starting and final values (in parts per million) of Au, K, and Ge noted in Tables 5, 6, and 7 in order to evaluate this theory.

Table 5. Values for Au

	Starting Value*	Final Value**
Test 1	<0.5	5540
Test 2	<0.5	174
Test 3	<0.5	5
Test 4	<0.5	7
Test 5	<0.51	252
Tests 6–10	<0.51	<0.01

*Certificate of analysis of copper electrodes
**ICP analysis from New Hampshire Materials Laboratory

Table 6. Values for K

	Starting Value*	Final Value**
Test 1	36.6	349
Test 2	36.6	750
Test 3	36.6	3
Test 4	36.6	28
Test 5	<1.615	181
Test 6	<1.615	6.09
Test 7	<1.615	6.50
Test 8	<1.615	23.27
Test 9	<1.615	11.66
Test 10	<1.615	4.41

*Certificate of analysis of lithium pieces, copper electrodes, and quartz tube
**ICP analysis from New Hampshire Materials Laboratory

The gap between starting value and final value in several of the experiments brings the above "concentration" theory into question. It also seems to make prior contamination only a remote possibility.

Table 7. Values for Ge

	Starting Value*	Final Value**
Test 1	<0.02	416
Test 2	<0.02	3596
Test 3	<0.02	252
Test 4	<0.02	44
Test 5	<0.02	<0.01
Tests 6–10	<0.02	<0.01

*Certificate of analysis of copper electrodes.
**ICP analysis from New Hampshire Materials Laboratory

In Test 1, for example, the starting value of Au in the copper electrode is listed as <0.5 ppm., a hypothetical value below the detection limit. The final value reported by the independent laboratory for Au is 5540 ppm; an apparent 10k-fold increase over the hypothetical starting value (Table 5). Similarly, in Test 2, the hypothetical starting value of Ge in the copper electrode is listed as <0.02 ppm, while the final value reported by independent analysis is indicated as 3596 ppm, an apparent 175k-fold increase (Table 7).

The possibility we propose, and that wish to pursue, is that of low energy nuclear reaction, so that under the conditions of electrified vacuum:

1. The final value of K is produced by low energy fusion:
   $^7Li + ^{32}S \rightarrow ^{39}K$
   Lithium-7 + sulfur-32 → potassium-39

2. The final value of Ge is produced by low energy fusion:
   $^7Li + ^{63}Cu \rightarrow ^{70}Ge$
   Lithium-7 + copper-63 → germanium-70

3. The final value of Au is produced by low energy fission:
   $^{204}Pb \rightarrow ^7Li + ^{197}Au$
   Lead-204 → lithium-7 + gold-197

The fission reaction is triggered by either of the above fusion reactions. However, results obtained from the experiments reveal a number of inconsistencies. For example, Tests 1–5 show consistent presence of Au, although with wide variation in the quantities detected. Tests 6–10 show no detectable Au in the final result. The values of Ge also vary greatly, as do the values of K. Why such discrepancy?

A partial answer may lie in the quality and intensity of the arc generated in the vacuum tube. Looking at color photos of the Test 1 (Fig. 3) and Test 5 (Fig. 7), we notice similarities in the intensity, focus, and color of the glow discharge in both tubes. The discharge appears to glow in a brilliant yellow-gold, deep red, and bluish purple color. It is curious that Au was detected in both experiments. In the 2012 tests (Fig. 8), the discharge seems to be more diffuse and lacking in the yellow-gold, bluish-purple, and deep red colors seen in the earlier tests. As we see in Table 5, Au was not detected in these tests, and only small amounts of K, much less than in the earlier tests, were noted (Table 6).

In my view, in Tests 6-10, the investigators failed to deliver sufficient power to the tube and/or achieve high enough temperatures to prompt the full set of reactions observed in Tests 1–5. Only weak reactions between lithium and sulfur, producing a fraction of the potassium detected in the earlier tests, were achieved. These weak low energy fusion reactions did not generate sufficient energy to trigger the fission reaction Pb → Li + Au. Hence, in contrast to Tests 1–5, in Tests 6–10, no Au, and only fractional amounts of K, were detected. Further research, in which tube pressure, variations in power and temperature, and other variables are carefully monitored and controlled, will be necessary to confirm these low energy reactions and reproduce them with reliability and precision.

Source: Edward Esko, "Ten Transmutation Experiments," *Infinite Energy*, No. 113, 2014.

Preliminary Research on Nuclear Remediation

Abstract

In my paper "LENR-Induced Transmutation of Nuclear Waste" (*Infinite Energy*, No. 104, 2012), I present possibilities for utilizing low energy nuclear reactions (LENR) as a means for converting nuclear materials into non-radioactive elements. The proposed experiments are based on LENR experiments conducted at the Quantum Rabbit (QR) laboratory since 2006. Papers on these experiments have appeared in *Infinite Energy* and have been complied in the book *Cool Fusion*, (Amber Waves, 2011.) In this paper I propose beginning the investigation into possible LENR nuclear remediation with experiments on the non-radioactive isotopes of strontium, iodine, and cesium. If these three non-radioactive isotopes can be successfully transmuted using low energy processes, it may be possible to apply this method to transmuting their radioactive isotopes, the nuclear fission products strontium-90, iodine-129, and cesium-137.

Transmutation of Strontium

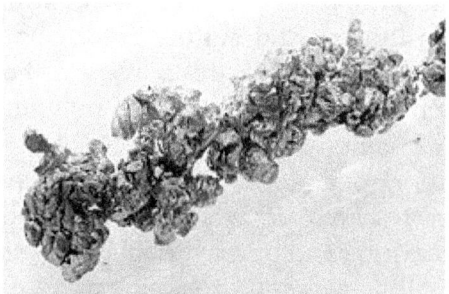

Fig. 1. Granules of pure strontium

Using low energy transmutation, it may be possible to convert non-radioactive strontium-88 into ruthenium-100. The proposed low energy fusion formula is as follows:

$$^{88}Sr + {}^{12}C \rightarrow {}^{100}Ru$$

Strontium-88 + carbon-12 → ruthenium-100

The proposed methodology for the test is essentially the same as that deployed in tests conducted by the research team at Quantum Rabbit lab in Owls Head, Maine (see "In Search of the Platinum Group Metals," *Infinite Energy,* No. 92, 2010). The test is conducted in a specially designed vacuum tube with electrodes of pure graphite. Once the graphite electrode (cathode) is secured in the tube a catalyst of pure sulfur is placed in the quartz midsection. A piece of strontium is placed in the center of the reaction zone. Then, a graphite anode is secured in the opposite end of the tube, so that it touches the strontium piece.

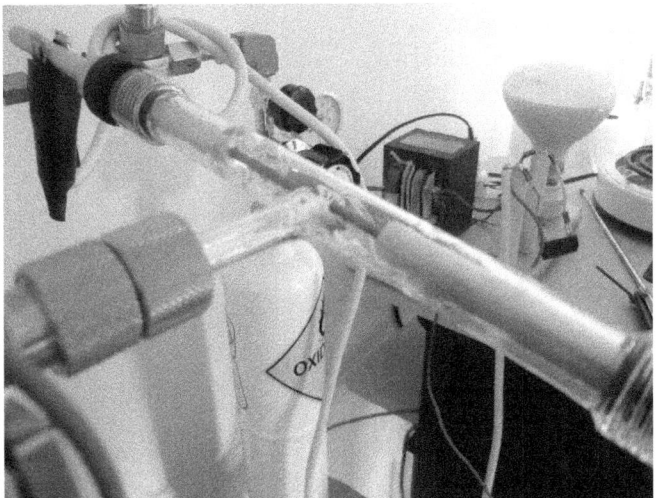

Fig. 2. Vacuum tube with graphite electrodes and strontium-sulfur test material

The tube is then pumped down to vacuum and backfilled with pure oxygen to about 3.5 torr. Direct current is passed through the electrodes, sufficient to spark plasma and vaporize the test materials. An electric heating tape or coil can be wrapped around the tube for supplemental baseline heat. Samples can be analyzed by ICP to test for the presence of ruthenium isotopes.

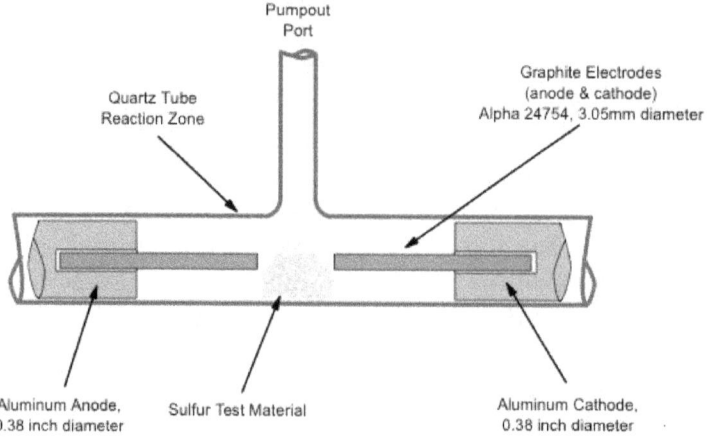

Fig. 3. In the proposed strontium test, strontium granules are mixed with the sulfur test material

Transmutation of Iodine

Fig. 4. Pure iodine crystals

The goal of the iodine study is to transmute the stable isotope of iodine (I-127) into tin (Sn-120) and barium (Ba-134) through low energy nuclear reactions as another step in demonstrating the possibility of the remediation of nuclear materials through LENR.

The first reaction is a low energy fission reaction:

$^{127}I \rightarrow {}^{7}Li + {}^{120}Sn$
Iodine-127 → lithium-7 and tin-120

The fission reaction is potentially triggered by the low energy fusion reactions:

$^7\text{Li} + {}^{32}\text{S} \rightarrow {}^{39}\text{K}$
Lithium-7 + sulfur-32 → potassium-39

$^{127}\text{I} + {}^7\text{Li} \rightarrow {}^{134}\text{Ba}$
Iodine-127 + lithium-7 → barium-134

This test can be conducted in the vertical tube described on page 52. In the test, a hole is drilled in the center of the lower copper electrode and a pellet of pure lithium is pressed into it. Iodine and sulfur test material are placed on top of the lower electrode. The glass/quartz tube is placed over the electrode assembly. The upper copper electrode is inserted into the tube and secured at the desired separation from the lower electrode and test materials. Oxygen or neon are backfilled to approximately 2 torr and plasma is struck using direct current.

Fig. 5. Edward Esko (left), Woody Johnson (center), and Alex Jack (right) monitor a vertical vacuum tube, 2012

The reaction zone can be heated with a hand torch or an electric heating tape or coil can be wrapped around the tube for supplemental baseline heat, with the goal being to achieve the relatively low melting temperatures of the elements utilized or sought after in the experiment: iodine, lithium, barium, and tin. Their melting temperatures are as follows:

Element	Melting Temperature °C
Iodine	114
Lithium	181
Tin	232
Barium	727

The experiment can continue until the reaction noticeably slows or the tube is in danger of breaking. Samples can be analyzed by ICP to test for the possible transmutation products tin, barium, and potassium. Further refinement of the testing process could demonstrate variations in isotopic composition of the transmutation products. Barium-134, the hypothetical transmutation product of iodine-127 and lithium-7 makes up only 2.4% of naturally occurring barium isotopes. If test samples were found to contain barium-134 in amounts significantly greater than this, it would help rule out contamination as the source of the barium and offer convincing proof of transmutation.

Transmutation of Cesium

Fig. 6. Cesium metal (left) and cesium chloride (right)

Cesium metal is highly reactive. In addition to igniting spontaneously in air, it reacts explosively with water even at low temperatures, more so than other members of the first group of the periodic table that includes lithium, sodium, and potassium. The reaction with solid water occurs at temperatures as low as −116 °C (−177 °F). Because of its high reactivity, the metal is classified as a hazardous material. It is stored and shipped in dry saturated hydrocarbons such as mineral oil.

Similarly, it must be handled under inert gas such as argon. It can be stored in vacuum-sealed borosilicate glass ampoules. (In quantities of more than about 100 grams, cesium is shipped in hermetically sealed, stainless steel containers.) In addition, cesium has a melting point of 28.4 °C (83.1 °F), making it one of the few pure metals which are liquid near room temperature. (Mercury is the only metal with a melting point lower than cesium.)

Due to these concerns, the team at Quantum Rabbit is not equipped to handle cesium in its elemental form. Accordingly, it recommended that research begin with one of the less volatile cesium compounds such as cesium chloride (CsCl.) Cesium chloride powder has a melting temperature of 646 °C, thus making it easier to handle than elemental cesium.

The goal of the cesium study is to transmute the stable isotope of cesium (Cs-133) into tellurium (Te-126), cerium (Ce-140), and neodymium (Nd-143 and Nd-144) through low energy fission and low energy fusion reactions.

The low energy fission reaction we propose testing is as follows:

$^{133}Cs \rightarrow {}^{7}Li + {}^{126}Te$
Cesium-133 → lithium-7 and tellurium-126

Our hope is to trigger the fission reaction by low energy fusion reactions:

$^{7}Li + {}^{32}S \rightarrow {}^{39}K$
Lithium-7 + sulfur-32 → potassium-39

$^{133}Cs + {}^{7}Li \rightarrow {}^{140}Ce$
Cesium-133 + lithium-7 → cerium-140

In a separate experiment, we propose testing the possibility of low energy fusion between cesium and boron:

$^{133}Cs + {}^{11}B \rightarrow {}^{144}Nd$
Cesium-133 + boron-11 → neodymium-144

These tests can be conducted in the vertical tube shown above. In the first test, a hole is drilled in the center of the lower copper electrode and a pellet of lithium metal is pressed into it. Sulfur crystals and cesium chloride power are sprinkled on the electrode surface. In the second test, cesium chloride powder and boron crystals are placed on the lower electrode surface. Following placement of test materials, the glass/quartz tube is placed over the electrode assembly. The upper copper electrode is inserted into the tube and secured at the desired separation from the lower electrode and test materials. Oxygen or neon are backfilled to approximately 2 torr and plasma is struck using direct current.

As in the previous tests, the reaction zone can be heated with a hand torch or an electric heating tape or coil can be wrapped around the tube for supplemental baseline heat. The experiment can continue until the reaction noticeably slows or the tube is in danger of breaking. Samples can be analyzed by ICP to test for the possible transmutation products tellurium, neodymium, cerium, and potassium.

Source: Edward Esko, "Preliminary Research on Nuclear Remediation, *Infinite Energy*, No. 110, 2013.

Appearance of Barium in Lithium-Iodine Plasma

Abstract

In studies conducted at Quantum Rabbit (QR) lab in Owls Head, Maine on April 10, 2013, funded by Woodland Energy and the New Energy Foundation, independent analysis of test samples by inductively coupled plasma mass spectroscopy (ICP-MS) revealed the anomalous presence of barium (Ba). Isotope distribution studies further revealed the presence of barium isotopes at variance with their natural distribution, notably elevated values for barium-134. The vacuum discharge tests employed copper (Cu) electrodes and lithium (Li) and iodine (I) test material. The appearance of novel isotope distribution in test samples diminishes the possibility that these results are due to contamination. Test results raise the possibility that barium was newly created through low energy transmutation.

Background

In my paper "Preliminary Research on Nuclear Remediation" (*Infinite Energy,* No. 110, 2013), I proposed beginning research into possible LENR nuclear remediation with tests on non-radioactive isotopes of strontium, iodine, and cesium, the concept being that if these non-radioactive isotopes are successfully transmuted, it may be possible to apply LENR to transmuting their radioactive isotopes. If scalable, low energy transmutation could offer a sustainable solution to the nuclear waste problem.

As I stated in my paper: "The goal of the iodine study is to transmute the stable isotope of iodine (I-127) into barium (Ba-134) through low energy nuclear reactions as another step in demonstrating the possibility of the remediation of nuclear materials through LENR." I proposed conducting tests on the following low energy fusion formula:

$$^{127}I + {}^{7}Li \rightarrow {}^{134}Ba$$
Iodine-127 + lithium-7 → barium-134

I went on to state: "Barium-134, the hypothetical transmutation product of iodine-127 and lithium-7 makes up 2.4% of naturally occurring barium isotopes. If test samples were found to contain barium-134 in amounts greater than this, it would help rule out contamination as the source of barium and offer proof of transmutation." In Test 2 conducted on April 10, barium-134 was found at 4.04%, nearly double the quantity found in naturally occurring barium.

The Lithium-Iodine Study
The procedure of the study was exceedingly simple. We conducted three experiments using the vertical vacuum tube deployed in previous tests (Fig. 2). Copper electrodes were inserted in the upper and lower ends of the tubes. A lithium plug was inserted in the center of the lower electrodes in Tests 1 and 2. Pure iodine crystals were placed on top the lithium inserts. In Test 3, for variation, no lithium plug was used. Instead a small piece of lithium was placed in the center of the lower copper electrode. Iodine crystals surrounded the lithium (Fig. 3).

Fig. 1. Lithium rod (left) and iodine crystals (right)

After pumping down to 3.5 torr, oxygen was admitted and the power turned on. When the arc was established and plasma struck, additional heat was provided by a hand-held torch. The arc was maintained for approximately 10 to 15 minutes, at which time the power was disconnected and the tubes allowed to cool.

Below is the worksheet I prepared prior to the experiments.

Lithium-Iodine Study
April 10, 2013

Inputs:
Copper electrodes
Lithium test material
Iodine test material
Oxygen fill

Procedures:
1. Insert Li plug in lower Cu electrode.
2. Place I on surface of Li plug (do not cover completely.)
3. Position electrode in tube.
4. Maneuver upper electrode into position.
5. Pump down to approx. 3.5 torr.
6. Admit O_2 fill and strike plasma
7. Apply torch as needed (optional.)
8. Continue approx. 15 minutes.

Note: In test 3, a piece of Li is placed on the lower electrode together with I, rather than inserting the Li plug in the electrode.

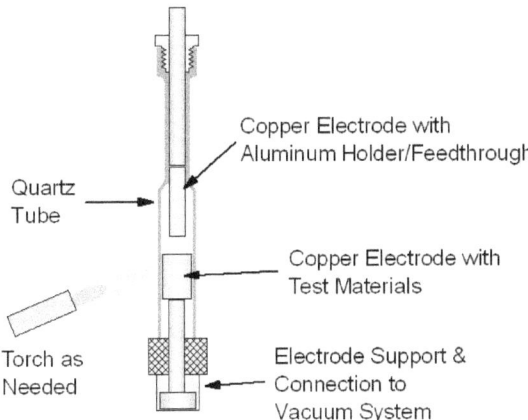

Fig. 2. The vertical tube used in the April 10 experiments

All three tests proceeded according to plan. We adjusted the electrode polarity between Test 1 and Tests 2 and 3. In Test 1, the upper electrode served as the anode, and in Tests 2 and 3, the upper electrode served as the cathode. The power supply was the same as that described in my earlier papers (see *Appendix A: The QR Power Supply*.)

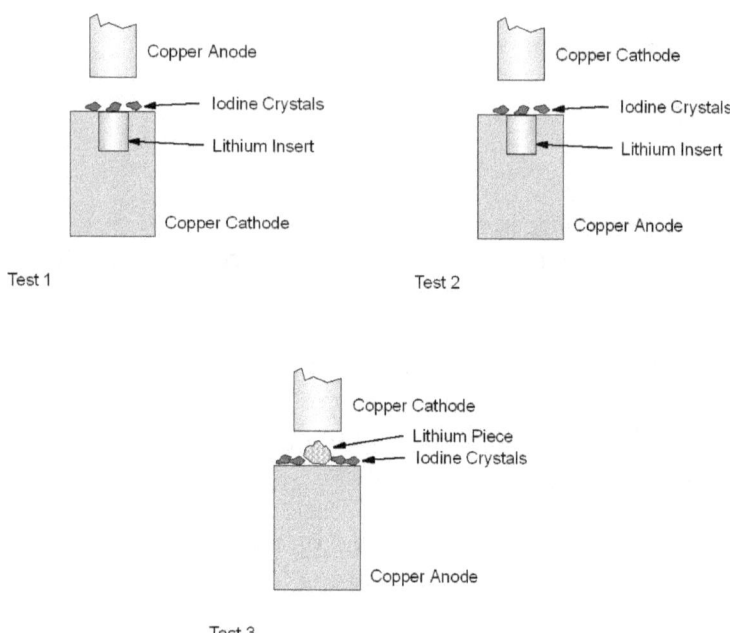

Fig. 3. Tests 1 and 2 (above) used a lithium insert. Test 3 used a piece of lithium surrounded by iodine crystals.

Three sets of samples were collected, labeled, and packed for shipping. The samples were sent to Northern Analytical Laboratory in Londonderry, NH with the following instructions:

Guidelines for Sample Analysis
April 2013

Sample Materials:

A. Test 1: One thick copper electrode (with lithium insert and lithium-iodine residue): one thin copper electrode; and one glass tube.

B. Test 2: One thick copper electrode (with lithium insert and lithium-iodine residue): one thin copper electrode; and one glass tube.
C. Test 3: One thick copper electrode with lithium-iodine residue; one thin copper electrode; and one glass tube.

Requested Analysis by ICPMS:
Barium (isotope distribution percentage)
Cesium

Procedure for Each Sample:
1. Scrape surface (top and sides) of Cu electrodes, including surface of Li insert.
2. Collect powder and flake residue (loose material) from plastic baggie.
3. Scrape residue from inner surface of tube.
4. Combine scrapings and loose material into one sample for analysis.

Note: Each sample to be analyzed separately.

Isotopes of Barium

Results came back on May 6. As predicted, barium was found in all three of the test samples, albeit in microscopic quantities. In Test 1 barium was found at 3.5 ppm, in Test 2 at 1.8 ppm, and in Test 3 at 463 ppm (Table 1). According to the Certificates of Analysis from Alfa Aesar, the supplier of the pure elements used in the test, barium was listed at <0.0005 ppm in the copper used for the electrodes. The lithium used in the test showed no value for barium.

Table 1. Values of Barium Before and After Experiment

Starting Concentration (ppm)*	Final Concentration (ppm)**
0.43	3.5 (Test 1)
0.43	1.8 (Test 2)
0.43	463 (Test 3)

*Source: Test Report from Northern Analytical Laboratory, 6/19/13
**Source: Test Report from Northern Analytical Laboratory, 5/6/13

Table 2. Iodine Certificate of Analysis

Product No.: 00158
Product: Iodine, crystalline, 99.99+% (metals basis)
Lot No.: D22X020

Purity	99.99%
Nonvolatile matter	0.004%
Chlorine and bromine (as Cl)	<0.005%

To provide a further control, we sent a sample of the iodine used in the tests to Northern Analytical Laboratory with instructions to test for barium. The control sample was taken from the same batch used in the experiments. The test results came back showing traces of barium in the iodine at 0.43 ppm, far below the amounts detected in the test samples. For example, in Test 1, the amount of barium was 8 times greater than that detected in the control, in Test 2, more than 4 times greater than the control, and in Test 3, barium was detected at a level more than 1,000 times greater than that detected in the control (Table 1).

The distribution of barium isotopes in the test samples varied somewhat from the distribution found in nature, suggesting the possibility that detected barium was newly created through low energy transmutation and not introduced through contamination. A comparison of the isotope distribution of natural barium and the barium found in the April experiments is shown in Table 3.

Table 3. Comparison of Barium Isotopes in Nature* and in the April 10 Experiments**

Isotope	Naturally Occurring %	Test 1 %	Test 2 %	Test 3 %
Ba132	0.101	1.36	3.92	0.19
Ba134	2.417	2.73	4.04	2.42
Ba135	6.592	6.62	6.18	6.72
Ba136	7.854	8.17	8.68	8.04
Ba137	11.232	10.9	10.9	11.9
Ba138	71.698	70.2	66.3	70.8

*Source: National Nuclear Data Center
**Source: Test Report from Northern Analytical Laboratory

The greatest variance in isotope distribution was seen in Test 2, with Ba-132 appearing at 3.92% compared to 0.101% in nature; Ba-134 at 4.04% compared to 2.417%; and Ba-138 at 66.3% compared to 71.698%.

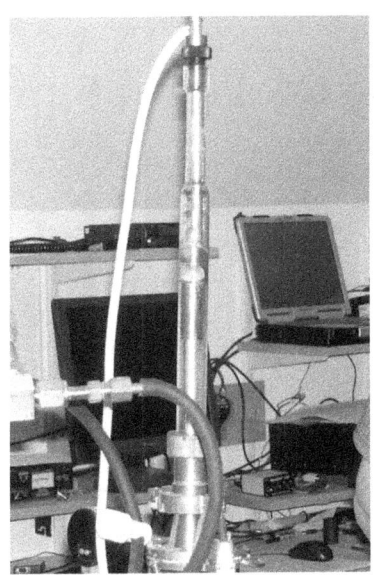
Fig. 4. Vertical tube used in the lithium-iodine study

Interpretation

Low energy fusion between iodine (atomic number 53) and lithium (atomic number 3) could explain the consistent appearance of barium (atomic number 56) in all three tests, with a peak at 463 ppm in Test 3. The distribution of barium isotopes is perhaps more difficult to understand. Ba-134 can be accounted for by the fusion of iodine-127 and lithium-7. That formula was predicted beforehand. Barium-132 may have arisen following the ejection of a neutron at the moment of fusion between Li-6 and I-127. Perhaps the low energy fusion process releases neutrons, some of which are captured by newly formed barium-134 nuclei, thus explaining the formation of isotopes heavier than Ba-134. More study is needed to fully explain the origin of the anomalous isotopes found in these experiments.

Source: Edward Esko, "Appearance of Barium in Lithium-Iodine Plasma," *Infinite Energy*, No. 111, 2013.

APPENDIX A
The QR Power Supply

The power supply used in the Quantum Rabbit (QR) vacuum discharge experiments is based on a microwave oven transformer. These transformers are current limited. With a 110 volt mains input the output will be about 3500 volts. As the transformer is loaded, the output voltage will become reduced to maintain a constant output current.

The "hot" output of the transformer is connected to a diode to provide a half-wave rectified output. Since the full output power of the transformer (about 1000 watts) could quickly overheat the QR experiment tube, the primary is ballasted with a high current series resistance consisting of a 600-watt heater element in parallel with a 150-watt incandescent lamp.

For monitoring the power supply we did not measure the actual tube voltage or current through the tube. Instead we monitored the primary voltage and the current into the primary. Voltage was measured with a digital volt meter (DVM) and the current with a commonly available "Kill-A-Watt" meter.

With no load on the transformer, the primary voltage would be close to the mains voltage (118 volts rms is normal here) with the current in the 3-4-amp range. With the tube in an arc condition (high current, low discharge voltage) the primary voltage would drop and the current would increase. The typical arc voltage would be in the 1-volt range vs. glow discharge in the range of a few hundred volts. Note that when the tube is conducting, the voltage across the tube will always be significantly lower than the approx. 3500-volt open circuit voltage. Multiplying the voltage times the current, about 300 V-A in most cases, may approximate power being delivered to the tube. **— Steve Hansen, QR Vacuum Consultant**

APPENDIX B
Inside the Cool Fusion Lab

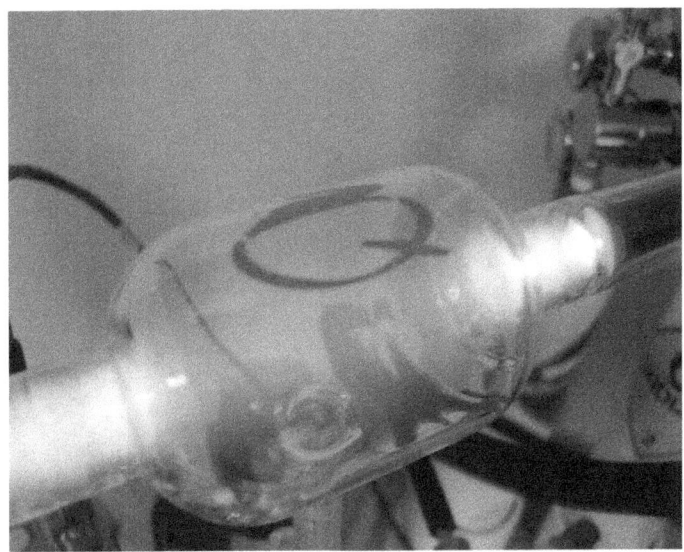

Helium plasma in QR vacuum tube, 2005

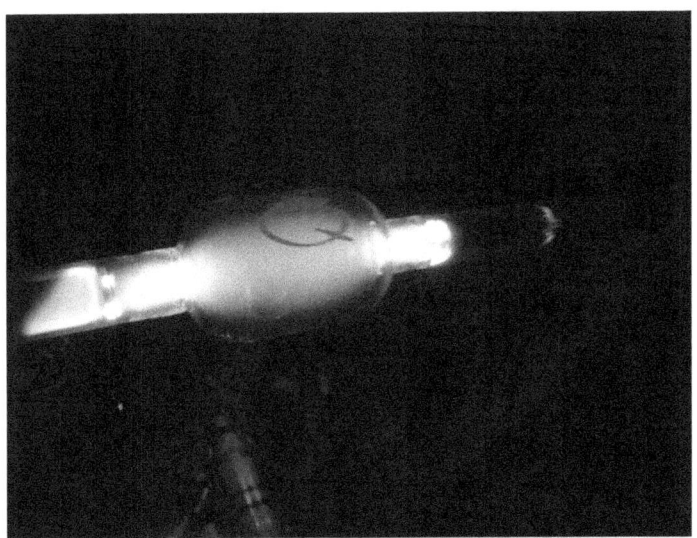

Oxygen plasma in QR vacuum tube, 2005

Edward Esko with QR vacuum tube, 2006

Lithium-argon plasma in QR tube, 2006

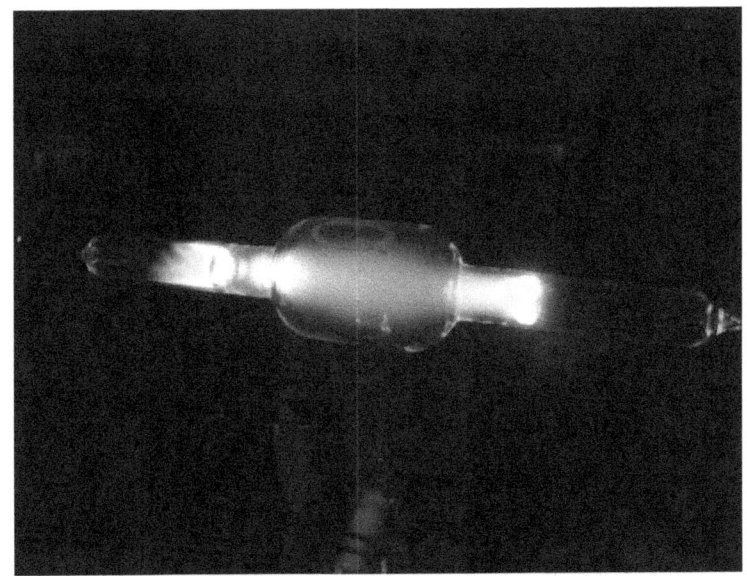
Krypton plasma in QR tube, 2006

Alex Jack adjusts lab worksheet, 2006

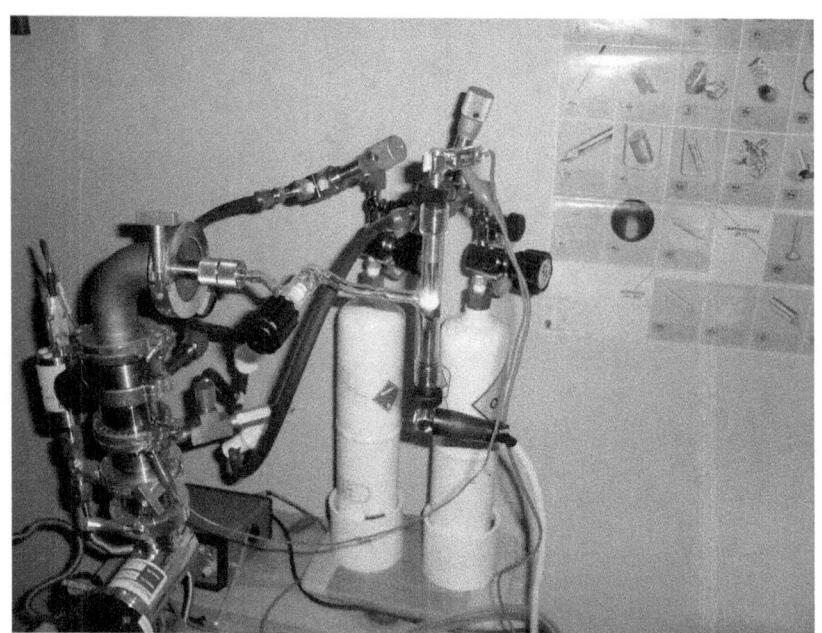
Sodium plasma in QR tube, 2006

Woody Johnson (left) with Alex Jack at the carbon-arc lab, 2007

Carbon-arc experiment using copper plate, 2007

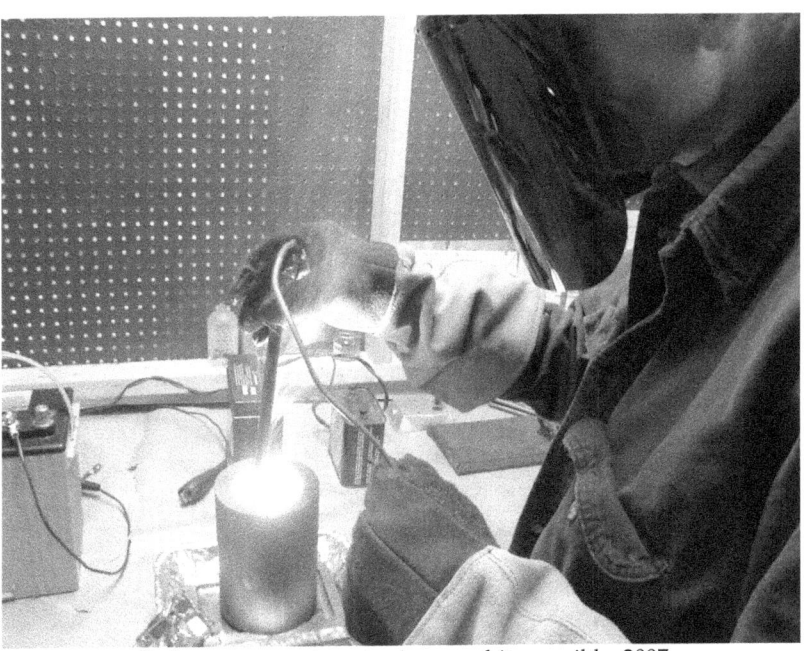
Carbon-arc experiment using graphite crucible, 2007

Lithium (center) in QR vacuum tube, 2008

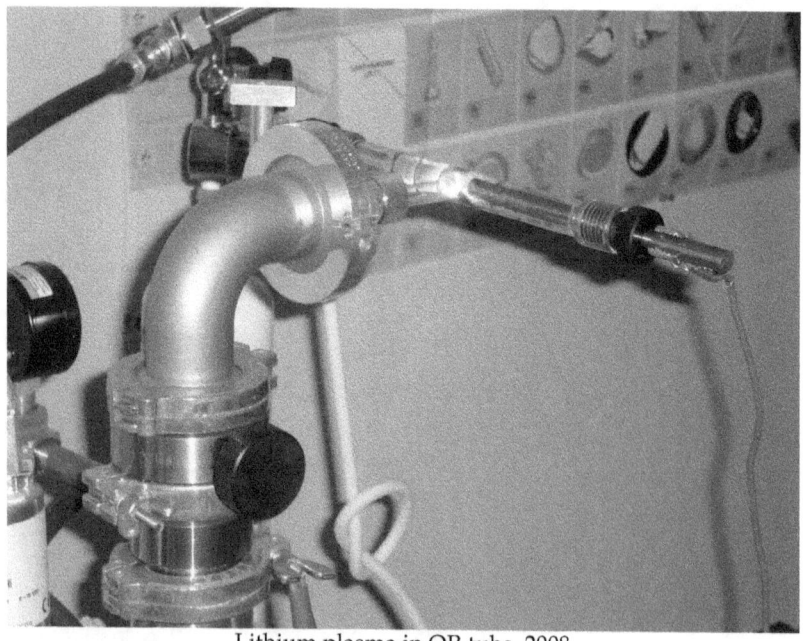
Lithium plasma in QR tube, 2008

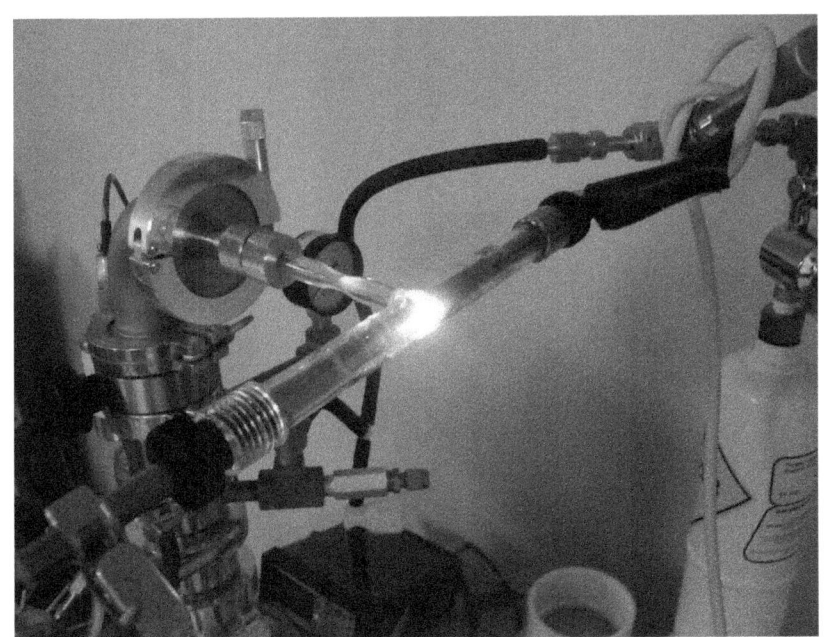
Sulfur plasma in QR tube, 2008

Bill Zebuhr monitors strontium-sulfur plasma, 2009

Woody Johnson observes lithium-sulfur plasma in QR tube, 2011

Lithium metal inside QR tube, 2012

Alex Jack monitors lithium-sulfur plasma in QR tube, 2012

Woody Johnson monitors lithium-sulfur plasma, 2012

Edward Esko monitors lithium-neon plasma in QR tube, 2012

Lithium metal and oxygen plasma in QR tube, 2012

RESOURCES

Quantum Rabbit LLC The Massachusetts Limited Liability Company (LLC) formed in 2005 by Edward Esko, Alex Jack, and Woody and Florence Johnson. For further information, including investment opportunities, contact Edward Esko, 109 Wendell Ave., Pittsfield MA 01201, (413) 442-1360, edwardesko@gmail.com.

CoolFusion.org Web site of Quantum Rabbit with information and articles on Cool Fusion.

Planetary Health Inc. A nonprofit educational organization and sponsor of Amberwaves, a grassroots network established to protect whole grains and other essential foods from genetic engineering, climate change, and other hazards; CERES, The Committee to Explore and Research Energy Solutions; and Amber Waves Press, publisher of *The Element Genome* and other literature. The quarterly *Amberwaves Journal* includes continuing coverage of QR experiments and activities and is $25/year.

For further information, including tax-deductible donations, contact Alex Jack, 305 Brooker Hill Road, Becket MA 01223, (413) 623-0012, shenwa@bcn.net, www.amberwaves.org.

Woodland Energy Co. A green energy company founded by Woodward Johnson and maker of the HUBERT® used in the QR experiments. For further information, contact Woodward Johnson, 200 Bush Hill Rd., Ashburnham MA 01430, (978) 827-5055, woodward1984@gmail.com, www.woodland-energy.com.

New Energy Foundation Sponsor of research in new energy science and technology, including QR experiments. NEF's magazine, *Infinite Energy*, reports on developments in cold fusion, cool fusion, and other promising technologies. Subscriptions are $29.95. For further information, contact: P.O. Box 2816, Concord NH 03302, www.infinite-energy.com.

ABOUT THE AUTHORS

Edward Esko began his study of macrobiotic cosmology, including low energy transmutation, in the 1970s with Michio Kushi in Boston. His study included the theory and practice of energy meridians, such as those used in acupuncture, the energetic effects of food, the universal stages of energy transformation, and the universal laws of harmony and balance as they appear in the world of physics and chemistry. In 1978 he traveled to Japan to deepen his understanding of Oriental cosmology and spirituality. He joined the faculty of Kushi Institute in 1980, and began lecturing there and around the world on practical applications of the cosmology of change as a means to achieve planetary health and peace. In 2004 Edward joined with Alex Jack and Woody Johnson to form Quantum Rabbit LLC, a Massachusetts Limited Liability Company, for the purpose of developing the theory of low energy transmutations presented by Louis Kervran, George Ohsawa, Michio Kushi, and other pioneers. Edward is the author of *Contemporary Macrobiotics*, *Yin Yang Primer*, and other books, and co-author, with Alex Jack, of *Cool Fusion*. Contact: edwardesko@gmail.com

Alex Jack served as a reporter in Vietnam, editor-in-chief of the *East West Journal*, general manager of the Kushi Institute, director of the One Peaceful World Society, and president of Planetary Health. He is the author or editor of many books, including *The Cancer Prevention Diet* with Michio Kushi, *Aveline Kushi's Complete Guide to Macrobiotic Cooking*, *The Mozart Effect* by Don Campbell, and *Buddha Standard Time* by Lama Surya Das. His multi-volume series, *Profiles in Oriental Diagnosis*, explores the creators of the modern mind, including Leonardo, Newton, Descartes, Darwin, and Pasteur. After the Chernobyl nuclear accident, he organized an airlift to the Soviet Union of foods that help protect against radiation. He has spoken at the Zen Temple in Beijing, the Cardiology Center in St. Petersburg, and Shakespeare's Globe Theatre in London. He is on the faculty of the Kushi Institute of America and a guest lecturer at the Kushi Institute of Europe and the Ohsawa Center in Japan. He lives in the Berkshires. Contact: shenwa@bcn.net.

www.ingramcontent.com/pod-product-compliance
Lightning Source LLC
Chambersburg PA
CBHW071751170526
45167CB00003B/998